Essential Differential Equations

Joseph P. Previte

Pennsylvania State University at Erie,

The Behrend College

ISBN 978-1-312-12695-4
Edition 3

Dedication

This is dedicated to the one of whom it is written:

"Worthy are you to take the scroll
and to open its seals,
for you were slain, and by your blood
you ransomed people for God
from every tribe and language
and people and nation,
and you have made them a kingdom
and priests to our God,
and they shall reign on the earth."

Revelation 5:9-10

Contents

Chapter 1

Introduction

1.1 Introduction to Differential Equations

Definitions

A *differential equation* is an equation involving an unknown function of one or more variables and at least one of its derivatives. If the unknown function only involves derivatives with respect to one variable, then the differential equation is called an *ordinary differential equation*, written as ODE for short. If the unknown function involves derivatives with respect to two or more variables, then the differential equation is called a *partial differential equation*, written as a PDE for short. The variables of the unknown function are called its *independent variables* and the unknown function's name is called the *dependent variable*.

The *order* of a differential equation is order of the highest order derivative that appears in the differential equation.

Example 1.1 *Classify each of the differential equations below as ordinary or partial. Also, identify the order, the unknown function's name, and its independent variables.*

(a) $u_{xx} + u_y = 1$

(b) $x\frac{dy}{dx} + y = 3xy$

(c) $z^{(3)} + 3z = 0$

Solution:

(a) This is a second order partial differential equation. The unknown function's name is u and its independent variables are x and y.
(b) This is a first order ordinary differential equation. The unknown function's name is y and its independent variable is x.
(c) This is a third order ordinary differential equation. The unknown function's name is z and its independent variable is unspecified. $\square$

A differential equation may be written using differential notation from calculus. In such cases, it is not necessarily clear which variables are independent or dependent as shown in the differential equation below:

$$(x^2 + y^2)\, dx + (y - x)\, dy = 0. \tag{1.1}$$

Here, Equation (1.1) can be rewritten as

$$(x^2 + y^2) + (y - x)\frac{dy}{dx} = 0,$$

where the function is y with independent variable x. Equivalently, Equation (1.1) can be rewritten as

$$(x^2 + y^2)\frac{dx}{dy} + (y - x) = 0,$$

where the function is x with independent variable y.

We say a function is a *solution* to (or solves) a differential equation if when the function and its derivatives are plugged into the differential equation, a true statement is obtained for all values of the independent variable that are in the function's domain.

Example 1.2 *Which of the following differential equations does $y = x^3$ solve?*
  (a) $\frac{dy}{dx} = y - x^3 + 3x^2$

  (b) $\frac{dy}{dx} = x^2$

  (c) $x\, dy + y^2\, dx = 0$

  (d) $\dfrac{dy}{dx} = \dfrac{3y}{x}$ *(Here, $x \neq 0$).*

Solution:
(a) For $y = x^3$ we have $\frac{dy}{dx} = 3x^2$. Plugging into the DE we obtain:

$$3x^2 = x^3 - x^3 + 3x^2$$

which is clearly true for all x so, yes, $y = x^3$ solves $\frac{dy}{dx} = y - x^3 + 3x^2$.

(b) Plugging into the DE $\dfrac{dy}{dx} = x^2$ we obtain:

$$3x^2 = x^2.$$

This statement is NOT true for all values of x in the domain of the function, so $y = x^3$ is not a solution to $\dfrac{dy}{dx} = x^2$.

(c) The DE $x\ dy + y^2\ dx = 0$ can be rewritten as $x\frac{dy}{dx} + y^2 = 0$. Plugging into the DE we obtain:

$$x(3x^2) + (x^3)^2 = 0$$

which is not true for all x so $y = x^3$ is not a solution to $x\ dy + y^2\ dx = 0$.

(d) Plugging into the DE we obtain

$$3x^2 = \frac{3x^3}{x}$$

which is true for all values of x except $x = 0$. Since the DE is not defined at $x = 0$, $y = x^3$ solves the DE whenever the DE makes sense, so $y = x^3$ is a solution to $\dfrac{dy}{dx} = \dfrac{3y}{x}$. $\square$

Note that in (d) above, a differential equation may not be defined for certain values of the variables. In such cases, we are entitled to ignore these values when searching for solutions. In other words, when considering the DE:

$$\frac{dy}{dx} = \frac{3y}{x},$$

it is clear that one does not need to consider when $x = 0$ since the DE does not give an equation for $x = 0$.

Example 1.3 *Consider the ODE $y'' + 3y' + 2y = 0$. Is $f(x) = c_1 e^{-x} + c_2 e^{-2x}$ a solution to this ODE, where c_1 and c_2 are constants?*

Solution: First compute:

$$\begin{aligned}
f'(x) &= -c_1 e^{-x} - 2c_2 e^{-2x} \\
f''(x) &= c_1 e^{-x} + 4c_2 e^{-2x}
\end{aligned}$$

Then substitute into the left-side of the ODE and simplify:

$$f''(x) + 3f'(x) + 2f(x)$$

$$= (c_1e^{-x} + 4c_2e^{-2x}) + 3(-c_1e^{-x} - 2c_2e^{-2x})$$
$$+ 2(c_1e^{-x} + c_2e^{-2x})$$

$$= c_1e^{-x} + 4c_2e^{-2x} - 3c_1e^{-x} - 6c_2e^{-2x}$$
$$+ 2c_1e^{-x} + 2c_2e^{-2x}$$

$$= 0$$

So, yes, $f(x) = c_1e^{-x} + c_2e^{-2x}$ is a solution to this ODE. $\square$

The above example illustrates that the function's name in a DE is simply a place-holder. In other words, the function $y = x^2$ is the same as the function $f(x) = x^2$ (or the function $z = x^2$), only the function's name has been is changed. Similarly, we will regard the DE $\frac{dy}{dx} = x$ as the same differential equation as $\frac{dy}{dt} = t$, since they only differ by a change in variable names.

Recall that functions can be defined *implicitly* from equations. For example,

$$xy^3 + y^2x^3 - 2 = x$$

defines y as an implicit function in terms of x (or vice versa). In calculus, we learned how implicit differentiation allows us to obtain the derivatives of implicit functions. Hence, a solution to a differential equation may also be given implicitly .

Example 1.4 *Show that*
$$xy^3 + y^2x^3 - 1 = x \tag{1.2}$$

gives an implicit solution to both:

(a) $\dfrac{dy}{dx} = \dfrac{1 - y^3 - 3x^2y^2}{3xy^2 + 2x^3y}$

(b) $\dfrac{dy}{dx} = \dfrac{xy^3 + y^2x^3 - x - y^3 - 3x^2y^2}{3xy^2 + 2x^3y}$

Solution:
(a) Using implicit differentiation, we differentiate both sides of (1.2) with respect to x treating y as a function of x:

$$\frac{d}{dx}\left(xy^3 + y^2x^3 - 1\right) = \frac{d}{dx}(x),$$

using the product rule,

$$1 \cdot y^3 + x(3y^2)\frac{dy}{dx} + 2y\frac{dy}{dx}x^3 + 3y^2x^2 = 1,$$

or

$$\frac{dy}{dx} = \frac{1 - y^3 - 3x^2y^2}{3xy^2 + 2x^3y},$$

which shows that equation (1.2) yields an implicit solution to the differential equation in (a).

(b) From equation (1.2),

$$1 = xy^3 + y^2x^3 - x,$$

substituting this expression into

$$\frac{dy}{dx} = \frac{1 - y^3 - 3x^2y^2}{3xy^2 + 2x^3y}$$

(from (a)), we obtain $\dfrac{dy}{dx} = \dfrac{xy^3 + y^2x^3 - x - y^3 - 3x^2y^2}{3xy^2 + 2x^3y}$. Thus, equation (1.2) yields an implicit solution to the differential equation in (b). $\square$

Writing Differential Equations

Oftentimes, for certain quantities of interest, the rate at in which the quantity changes is well-known as opposed to knowledge of the quantity itself. In situations such as these, differential equations naturally arise. For example, it is well-known that radioactive materials decay at a rate proportional to the amount of material present. Therefore, if $M(t)$ is the mass of a certain radioactive material at time t, then $\frac{dM}{dt} = -kM$ where $k > 0$ is a constant of proportion. (The actual value of k would be determined by how fast the material decays as well as the units with which time and mass are measured).

Example 1.5 *Money in an account grows at a rate proportional to the amount of money in the bank. If we assume that this growth is continuous, write a differential equation for $M(t)$ the amount of money in the bank at time t.*

Solution: $M(t)$ is the amount of money in the bank at time t so

$$M'(t) = kM(t),$$

where $k > 0$ is the constant of proportionality.   □

Example 1.6 *From physics, Newton's Law of Cooling (or Heating) states that the rate of change of the temperature of an object is proportional to the difference of the temperature T of that object and the ambient temperature of the room. Assuming the ambient temperature of the room is held constant at $M°$, write a differential equation for the rate of change of the temperature the object.*

Solution: Let $T(t)$ denote the temperature of the object at time t. Then by Newton's Law of Cooling:

$$T'(t) = k(T(t) - M)$$

or

$$\frac{dT}{dt} = k(T - M),$$

where k is a constant.   □

Example 1.7 *A 300 gallon tank contains is full of a mixture of alcohol and water. It is then drained at 3 gallons per minute. If the amount of alcohol at time t is $A(t)$ write an equation for $\frac{dA}{dt}$*

Solution: To do this problem, we need to realize that we are given information about $\frac{dA}{dt}$. Also, we realize that alcohol is leaving the tank, so $\frac{dA}{dt}$ should be negative.

Next, we consider the 3 gallons per minute that is leaving the tank. Is that all alcohol? Is it all water? We need to know know much of this mixture leaving is alcohol.

To find this, we need the amount of alcohol in 1 gallon of mixture at any time t. Note that the volume of the *mixture* is $V(t) = 300 - 3t$, since we lose one gallon each minute. Thus the amount of alcohol in one gallon must be

$$\frac{A(t)}{300 - 3t}.$$

Now we can write our differential equation since 3 gallons of mixture leave per minute

$$\frac{dA}{dt} = -3\left(\frac{A(t)}{300 - 3t}\right).$$

$\square$

First Order ODE From First Semester Calculus

The reader should recall that a fair portion of calculus involves solving differential equations. In particular if

$$\frac{dy}{dx} = f(x)$$

and $f(x)$ has antiderivative $F(x)$ then $y = F(x) + C$. So finding an antiderivative amounts to solving a differential equation.

Example 1.8 *Find all solutions of* $\dfrac{dy}{dx} = x^2$

Solution: By integrating, we obtain

$$y = \frac{1}{3}x^3 + C$$

where C is any constant. $\square$

It is at this point that we raise a technical concern that is perhaps overlooked in a standard calculus class, namely: How do we know that we have ALL solutions to this ODE? In other words, if $g(x)$ is also a solution to $\dfrac{dy}{dx} = x^2$, why must $g(x) = f(x) + C$ for some constant C? To address this issue, assume that $g(x)$ is a solution $\dfrac{dy}{dx} = x^2$. Then

$$h(x) = g(x) - \frac{1}{3}x^3$$

differentiating h we obtain $h'(x) = 0$ for all x which implies that $h(x)$ is a constant function, so there must be a C so that $g(x) = \frac{1}{3}x^3 + C$.

When all solutions to a differential equation have been found, we say that we have obtained the *general solution* to the ODE.

A first differential equation together with one specified condition that a solution must satisfy (called an *initial condition*) is called an *initial value problem*, or IVP for short . A solution to a particular IVP is called a *particular solution.*

Example 1.9 *Find the solution to initial value problem* $\dfrac{dy}{dx} = x^2;\ y(1) = 2$

Solution: The general solution to the DE is

$$y = \frac{1}{3}x^3 + C$$

where C is any constant. Since we want the particular solution that satisfied $y(1) = 2$ (i.e., when $x = 1$ we need $y = 2$), we plug into the general solution to obtain an equation for C. Namely,

$$2 = \frac{1}{3} + C$$

or $C = \frac{5}{3}$. Thus the particular solution to the IVP is

$$y = \frac{1}{3}x^3 + \frac{5}{3}. \ \square$$

Exercises

In each of 1-6, classify each of the differential equations below as ordinary or partial. Also, identify the order, the unknown function's name, and its independent variable(s).

1. $\dfrac{dz}{dt} + z = t$

2. $u_{xy} + 2u_{xxz} = u$

3. $\dfrac{d^3y}{dx^3} + \left(\dfrac{d^2y}{dx^2}\right) \cdot \left(\dfrac{dy}{dx}\right) = 2xy$

4. $x\,dx + y\,dy = 0$

5. $z_x = z_y$

6. $x^2y'' + \frac{1}{4}y = 0$

In 7-10, determine whether $y = \sqrt{x}$ solves the following DE. Show all work.

7. $\dfrac{dy}{dx} = \dfrac{1}{2y}$

8. $x\,dy + y\,dx = 0$

9. $x^2y'' + \frac{1}{4}y = 0$

10. $x\,dx + y\,dy = 0$

11. Determine whether $x^3 + y^4 = xy$ gives an implicit solution to

    (a) $\dfrac{dy}{dx} = \dfrac{3x^2 - y}{4y^3 - x}$

    (b) $(3x^2 - y)\,dx + (4y^3 - x)\,dy = 0$

    (c) $\dfrac{dy}{dx} = \dfrac{y^4 - 2x^3}{4y^3x - x^2}$

12. The rate of population growth of a certain country grows at a rate proportional to the population itself. If $P(t)$ is the population at time t write a differential equation that describes the population change.

13. $M(t)$ kg of salt is dissolved in a tank that holds $400 - 2t$ liters of saltwater at time t, where t is measured in minutes. The mixture is drained at 2 liters per minute. Write a differential equation that describes that $\frac{dM}{dt}$, the rate of change in the amount of salt in the tank at time t.

14. A colony of ants spreads at a rate proportional to the square root of the area that has already been colonized. If $A(t)$ is the current area of the colony, write a differential equation that describes the rate of change of the area of the colony.

15. From physics, it can be assumed that a free falling object near the surface of the earth falls with constant acceleration. If $s(t)$ is the height of a free-falling object at time t, write a differential equation that describes the acceleration.

16. Solve the following initial value problems

    (a) $\dfrac{dy}{dx} = \dfrac{1}{x^2 + 4}$; $y(1) = \pi$

    (b) $\dfrac{dy}{dt} = te^t$; $y(0) = 1$

    (c) $\dfrac{dy}{dx} = 2y$; $y(0) = 1$ (Hint: Use $\frac{dx}{dy} = \frac{1}{\frac{dy}{dx}}$ and solve for x)

17. A 100 gallon tank contains is full of a mixture of salt and water. It is then drained at 8 gallons per minute. If the amount of salt at time t is $S(t)$ pounds of salt, write an equation for $\frac{dS}{dt}$. Hint: Find the salt in 1 gallon of mixture.

Chapter 2

Methods for Solving First Order ODEs

2.1 Separable First Order ODE

> **Separable First Order ODE**
> A first differential equation is called *separable* if it can be rewritten in the form
>
> $$g(y) \, dy = f(x) \, dx \qquad (2.1)$$
>
> where $g(y)$ is a function of y (which includes the case of $g(y)$ equalling a constant) and $f(x)$ is a function of x (which includes the case $f(x)$ equalling a constant).

Note that if a differential equation can be rewritten in the form of equation (2.1) by changing the names of the variables, then it is separable. Not all first order differential equations are separable. The reader should verify that $\frac{dy}{dx} = x + y$ is not separable.

Solving a separable first order differential equation amounts to integrating both sides with respect to their respective variables as shown below.

11

Solving Separable First Order ODE

Suppose that $G(y)$ is an antiderivative of $g(y)$ and that $F(x)$ is an antiderivative of $f(x)$. Then

$$G(y) = F(x) + C$$

solves the separable differential equation

$$g(y) \, dy = f(x) \, dx$$

Proof: By implicitly differentiating

$$G(y) = F(x) + C$$

with respect to x we get

$$G'(y)\frac{dy}{dx} = F'(x)$$

or

$$g(y) \, dy = f(x) \, dx \quad \square$$

The method to solve a separable ODE is given below:

Method for Solving Separable First Order ODE

1. Rewrite the DE in differential notation with the variables separated on either side of the equation:

$$f(x) \, dx = g(y) \, dy$$

(This is only possible if the DE is separable!)

2. Integrate both sides with respect to their respective variables.

$$\int f(x) \, dx = \int g(y) \, dy + C$$

(Use only one constant of integration, as shown)

3. When possible, solve the implicit solution in the previous step for y (or for whatever happens to be the dependent variable).

Example 2.1 *Find all solutions to* $\dfrac{dy}{dx} = 3y$

Solution: We rewrite the DE in differential notation and note that it is separable (with $g(y) = \frac{1}{y}$ and $f(x) = 3$).

$$\frac{1}{y}\, dy = 3\, dx$$

(Note that this is only valid, so long as $y \neq 0$). Next, we antidifferentiate both sides:

$$\int \frac{1}{y}\, dy = \int 3\, dx$$

and obtain

$$\ln|y| = 3x + C,$$

which is an implicit solution to $\dfrac{dy}{dx} = 3y$, where C is an arbitrary constant.

This solution can be solved explicitly for y as

$$|y| = e^{3x+C}$$

so

$$|y| = e^{C} e^{3x}$$

or

$$|y| = K e^{3x},$$

where $K = e^{C}$ is an arbitrary positive constant. Solving for y we obtain

$$y = \pm K e^{3x}.$$

Noting that $\pm K$ is an arbitrary non-zero constant, we relabel it as K, dropping the condition that K be positive, or

$$y = K e^{3x}.$$

We can see by simply plugging into the DE that the constant function $y = 0$ also solves the DE, so

$$y = K e^{3x},$$

where K is any arbitrary constant yields a solution to the DE. $\qquad\square$

Note 1: In the previous example, when writing the DE in differential notation, we saw that $y \neq 0$, but in fact we saw that the constant function $y = 0$ itself solved the DE. This is the case in general, for if $y = c$ is a zero of $g(y)$ (i.e., $\frac{1}{g(c)}$ is undefined) then $y = c$ solves $g(y) \, dy = f(x) \, dx$.

Note 2: In the previous example, after antidifferentiating both sides, only ONE constant of integration is required (why?).

Example 2.2 *Find a solution to the initial value problem* $\dfrac{dy}{dx} = xy^2$; $y(1) = 2$

Solution: We rewrite the DE in differential notation and note that it is separable.

$$\frac{1}{y^2} \, dy = x \, dx$$

(Note that this is only valid, so long as $y \neq 0$). Next, we antidifferentiate both sides:

$$\int \frac{1}{y^2} \, dy = \int x \, dx$$

and obtain

$$-y^{-1} = \frac{1}{2}x^2 + C$$

Plugging in the initial condition, we obtain

$$-\frac{1}{2} = \frac{1}{2}1^2 + C$$

or

$$C = -1.$$

So

$$-y^{-1} = \frac{1}{2}x^2 - 1$$

solves the initial value problem. This can be solved explicitly for y as

$$y = \frac{1}{1 - \frac{1}{2}x^2} \qquad \square$$

Note that in the above example, we could have solved explicitly for y first, then obtained C.

Changing Variables to a Separable Equation

Often in mathematics, a change of variables can be used to transform a problem into one that can more readily be solved. We provide a few examples that can be solved by a change of variables.

Example 2.3 *Solve the DE*

$$\frac{dy}{dx} = \sin(\frac{y}{x}) + \frac{y}{x}$$

Solution: Note that the DE is not separable. Consider the change of variable $v = \frac{y}{x}$ (or $vx = y$).

Differentiating $vx = y$ with respect to x we see that

$$v + \frac{dv}{dx}x = \frac{dy}{dx}$$

So by substitution:

$$v + \frac{dv}{dx}x = \sin(v) + v$$

We obtain

$$\frac{dv}{dx}x = \sin(v),$$

which is separable.

$$\csc(v)\, dv = \frac{1}{x}\, dx,$$

and

$$-\ln|\csc v + \cot v| = \ln|x| + C$$

So an implicit solution to the original DE is given by

$$-\ln\left|\csc(\frac{y}{x}) + \cot(\frac{y}{x})\right| = \ln|x| + C \quad \square$$

In general, this technique works if $\frac{dy}{dx}$ can be expressed as a function of $\frac{y}{x}$. In Example 2.3, this function was $F(v) = \sin v + v$.

Exercises

Solve each of the following:

1. $\dfrac{dz}{dt} = zt$

2. $\dfrac{dy}{dx} = \dfrac{x}{y}$

3. $\dfrac{dy}{dx} = (y^2 + 1)x$

4. $\dfrac{dy}{dx} = y(y + 1)$

5. $z' = z^2$

6. $\dfrac{dz}{dt} = t \sin z$

7. $x^2 y \, dx + y^3 x \, dy = 0$

8. $\dfrac{dy}{dx} = \ln(\dfrac{y}{x}) + \dfrac{y}{x}$

9. $\dfrac{dz}{dx} = \left(\dfrac{z}{x}\right)^2 + 2\left(\dfrac{z}{x}\right)$

Find solutions for the following initial value problems

10. $\dfrac{dy}{dx} = \sqrt{y}; \quad y(1) = 4$

11. $\dfrac{dy}{dx} = 3yx; \quad y(1) = 2$

12. $\dfrac{dy}{dx} = 3yx; \quad y(1) = 0$

13. $\theta \, dr = d\theta; \quad r(\pi) = 1$

14. $y \, dy - 2x \csc y^2 \, dx = 0; \quad y(1) = \sqrt{\dfrac{\pi}{2}}$

15. $\dfrac{dy}{dx} = \left(\dfrac{y}{x}\right)^2 + \dfrac{y}{x}$; $y(1) = 2$

16. A population of *E. coli* bacteria doubles in roughly 20 minutes. If a model for the population is $\dfrac{dP}{dt} = kP$, find k, where t is measured in days. (Hint: Solve the DE and use $P(\frac{20}{1440}) = 2P(0)$ to find k.)

2.2 First Order Linear ODE

A differential equation is called a *linear* first order ODE if it can be rewritten into the form:

$$a_1(x)\frac{dy}{dx} + a_0(x)y = g(x). \tag{2.2}$$

Here $a_1(x)$, $a_0(x)$ and $g(x)$ are functions of x. So long as $a_1(x) \neq 0$, by dividing, we can write any first order linear differential equation in *standard form*, shown below.

Standard Form of a First Order Linear ODE

A first order linear ODE is said to be in standard form if it is in the form

$$\frac{dy}{dx} + P(x)y = Q(x) \tag{2.3}$$

for functions $P(x)$ and $Q(x)$.

Any first order linear has a solution given below:

Solving a First Order Linear ODE in Standard Form

Consider the differential equation

$$\frac{dy}{dx} + P(x)y = Q(x).$$

If $P(x)$ has antiderivative $\int P$

$$y = \frac{\int Q(x)e^{\int P}\, dx + C}{e^{\int P}}.$$

(Here, C is the constant of integration of the outer integral in the numerator).

Proof: Use the quotient rule to differentiate

$$y = \frac{\int Q(x)e^{\int P}\, dx + C}{e^{\int P}}$$

with respect to x and the fact that $\frac{d}{dx}\int P = P(x)$ obtain

$$\frac{dy}{dx} = \frac{\left(Q(x)e^{\int P}\right)e^{\int P} - \left(\int Q(x)e^{\int P}\,dx + C\right)e^{\int P}P(x)}{\left(e^{\int P}\right)^2}$$

The right hand side simplifies to

$$= Q(x) - P(x)\left(\frac{\int Q(x)e^{\int P}\,dx + C}{e^{\int P}}\right) = Q(x) - P(x)y \quad \square$$

Example 2.4 *Solve the DE*

$$\frac{dz}{dx} = 2\frac{z}{x} + x; \quad x > 0$$

Solution: We first write the DE in standard form:

$$\frac{dz}{dx} - \frac{2}{x}z = x.$$

We then identify $P(x) = -\frac{2}{x}$ and $Q(x) = x$. Next, compute $\int P = -2\ln x$ (note that only one antiderivative is required and since $x > 0$, we do not write $\ln|x|$).

Next we compute $e^{\int P}$:

$$e^{\int P} = e^{-2\ln x} = e^{\ln x^{-2}} = \frac{1}{x^2}.$$

Substituting into the formula:

$$y = \frac{\int Q(x)e^{\int P}\,dx + C}{e^{\int P}},$$

we obtain

$$y = \frac{\int x \cdot \frac{1}{x^2}\,dx + C}{\frac{1}{x^2}}$$

and finally get

$$y = \frac{\ln x + C}{\frac{1}{x^2}} = x^2\ln x + Cx^2. \quad \square$$

For a first order linear differential equation in standard form, the expression $\exp^{\int P}$ is called the *integrating factor*. Instead of simply memorizing the above formula, an alternate way to solve equations of the form (2.3) is to multiply equation (2.3) by the integrating factor $\exp^{\int P}$ and then realizing that the resulting left-hand side is equal to $\frac{d}{dx}\left(y \cdot \exp^{\int P}\right)$ by the product rule.

Example 2.5 *Solve the DE*

$$\frac{dy}{dx} = x - 3y$$

using the alternate method described above.

Solution: We first write the DE in standard form:

$$\frac{dy}{dx} + 3y = x.$$

We then identify $P(x) = 3$ and $Q(x) = x$. Multiplying both sides of the differential equation by the integrating factor $e^{\int P} = e^{3x}$ we obtain:

$$e^{3x}\frac{dy}{dx} + 3e^{3x}y = xe^{3x}.$$

Note that by the product rule (and chain rule), the left-hand side of the above expression is simply $\frac{d}{dx}\left(e^{3x}y\right)$

Rewriting, we obtain

$$\frac{d}{dx}\left(e^{3x}y\right) = xe^{3x}.$$

We integrate both sides with respect to x and obtain

$$e^{3x}y = \int xe^{3x}\,dx$$

The right-hand side (after integration by Parts) simplifies to

$$\frac{1}{3}xe^{3x} - \frac{1}{3}e^{3x} + C$$

Hence,

$$e^{3x}y = \frac{1}{3}xe^{3x} - \frac{1}{3}e^{3x} + C$$

and solving for y, we obtain

$$y = \frac{1}{3}x - \frac{1}{3} + Ce^{-3x} \quad \square$$

Note: Sometimes $P(x)$ or $e^{\int P(u)\,du}Q(x)$ cannot be integrated: then the solution to $\frac{dy}{dx} + P(x)y = Q(x)$

(expressed as a definite integral) is

$$y(x) = \frac{\int_{x_0}^{x} Q(u)e^{\int P} \, du + C}{e^{\int P}}$$

where $\int P$ is an antiderivative of P with no constant of integration, and is expressed in terms of independent variable u in the numerator and x in the denominator.

Exercises

Solve the following DE:

1. $\dfrac{dy}{dx} - \dfrac{1}{x}y = x^3; \quad x > 0$

2. $\dfrac{dy}{dx} - y = e^{2x}$

3. $\dfrac{dz}{dt} = t + z$

4. $x\, dy = \left(x^3 - y\right) dx$

5. $\dfrac{dz}{dt} = \cos t - z \cot t; \quad 0 < t < \dfrac{\pi}{2}$

6. $\dfrac{dy}{dx} = \dfrac{y}{2y - x}$ (Hint: solve for x in terms of y)

Solve the following initial value problems

7. $\dfrac{dy}{dx} + \dfrac{3}{x}y = 1; \quad y(1) = 2$

8. $\dfrac{dy}{dx} + y = 2; \quad y(0) = -1$

9. $\dfrac{dz}{dx} + 2xz = x; \quad z(1) = 2$

10. The population of deer on a certain island is modeled by $\dfrac{dP}{dt} = \dfrac{3}{4}P - 100$.

    (a) Solve the DE for any initial condition $P(0)$.

    (b) If $P(0) = 45$ explain what will happen to the population. What if $P(0) = 200$?

11. Money in a bank account at time t can be modeled by

    $\frac{dy}{dt} = ry + D$ where the constant r is related to the interest rate and the constant D is the related to regular deposits at a constant rate.

    (a) Solve the DE in general (leave r and D constant).

(b) Solve for C (the constant of integration) in terms of the initial amount in the bank $y(0)$

12. Solve the IVP $\frac{dy}{dx} = \frac{1}{x}y + x^2$, $y(1) = 9$

13. Solve $\dfrac{dy}{dx} = y + x$

14. The following DE is linear and has integrals that cannot be simplified. However, one can use technology to estimate the definite integrals. Use technology to plot the solution for $0 \leq x \leq 1$:

$$\frac{dy}{dx} - xy = 1; \quad y(0) = 1$$

15. The following DE is linear and has integrals that cannot be simplified. However, one can use technology to estimate the definite integrals. Use technology to plot the solution $0 \leq x \leq 1$:

$$\frac{dy}{dx} - \sin(x^2)y = 1; \quad y(0) = 1$$

2.3   Exact Differential Equations

A differential equation is called *exact* when it is written in the specific form

$$F_x \, dx + F_y \, dy = 0, \tag{2.4}$$

for some continuously differentiable function of two variables $F(x, y)$. (Note that in the above expressions $F_x = \frac{\partial F}{\partial x}$ and $F_y = \frac{\partial F}{\partial y}$).

The solution to equation (2.4) is given implicitly by

$$F(x, y) = C. \tag{2.5}$$

We see this by implicitly differentiating

$$F(x, y) = C.$$

with respect to x (and using the chain rule from multivariable calculus) we see that an exact differential equation must be of the form:

$$F_x + F_y \frac{dy}{dx} = 0,$$

which can be written as

$$F_x \, dx + F_y \, dy = 0.$$

Example 2.6 *Find the exact differential equation that is solved by*

$$x^2 y + y^3 \sin x = C$$

Solution: Differentiating, we obtain

$$\left(2xy + y^3 \cos x\right) dx + \left(x^2 + 3y^2 \sin x\right) dy = 0 \tag{2.6}$$

Note that one needs to be extremely careful calling a differential equation exact, since performing algebra on an exact differential equation can make it no longer exact. In other words, the differential equation

$$\left(2xy^2 + y^4 \cos x\right) dx + \left(yx^2 + 3y^3 \sin x\right) dy = 0$$

is algebraically equivalent to equation (2.6) but it is not exact, even though it is still solved by

$$x^2 y + y^3 \sin x = C. \square$$

One should recall that if F is continuously differentiable then the mixed partial derivatives of F must match namely, $F_{xy} = F_{yx}$. This gives us a method to detect if a differential equation is exact namely:

Exactness Test and Method to Solve an Exact DE

Consider the differential equation

$$M(x,y) \ dx + N(x,y) \ dy = 0$$

where M and N are both continuously differentiable functions with continuous partials M_y and N_x. If $M_y = N_x$, then the DE is exact. The implicit solutions are given by $F(x,y) = C$ where $F = \int M \ dx$ and $F = \int N \ dy$, simultaneously, up to a constant C.

We first show that one can obtain a function so that $F = \int M \ dx = \int N \ dy$, simultaneously, up to a constant C. Given that $M_y = N_x$. Consider $\int M \ dx - \int N \ dy$. Rewrite this as:

$$\int (\int M_y \ dy) \ dx - \int (\int N_x \ dx) \ dy,$$

which equals

$$\int \int 0 \ dx \ dy$$

which is a constant.

Suppose that such an F now exists so that $F = \int M \ dx$ and $F = \int N \ dy$, simultaneously. Then differentiating we obtain

$$F_x \ dx + F_y \ dy = 0, \tag{2.7}$$

or

$$M \ dx + N \ dy = 0. \tag{2.8}$$

Moreover, since $F_{xy} = F_{yx}$ we must have $M_y = N_x$. $\qquad \square$

Example 2.7 *Use the test for exactness to show that the DE is exact, then solve it.*

$$\left(x^2 + xy - y^2\right) \ dx + \left(\frac{1}{2}x^2 - 2xy\right) \ dy = 0. \tag{2.9}$$

Solution:

In this problem, $M = x^2 + xy - y^2$ and $N = \frac{1}{2}x^2 - 2xy$. Thus,

$$M_y = x - 2y$$

and

$$N_x = x - 2y,$$

which implies that the differential equation is exact.

To obtain F we compute $F = \int M \ dx$ and $F = \int N \ dy$.

$$F = \int M \ dx = \int x^2 + xy - y^2 \ dx = \frac{1}{3}x^3 + \frac{1}{2}x^2 y - xy^2 + h_1(y)$$

where $h_1(y)$ is an unknown function of y. Similarly,

$$F = \int N \ dy = \int 12x^2 - 2xy \ dy = \frac{1}{2}x^2 y - xy^2 + h_2(x)$$

where $h_2(x)$ is an unknown function of x.

For F to equal both simultaneously, we must have $h_2(x) = \frac{1}{3}x^3$ and $h_1(y) = 0$. Thus $F(x, y) = \frac{1}{3}x^3 + \frac{1}{2}x^2 y - xy^2$ and hence,

$$\frac{1}{3}x^3 + \frac{1}{2}x^2 y - xy^2 = C$$

is the solution to the DE.   □

Example 2.8 *Use the test for exactness to show that the DE is exact, then solve the initial value problem.*

$$(ye^{xy}) \ dx + (xe^{xy} + \sin y) \ dy = 0 \quad y(0) = \pi \qquad (2.10)$$

Solution:

In this problem, $M = ye^{xy}$ and $N = xe^{xy} + \sin y$. Thus,

$$M_y = e^{xy} + xye^{xy}$$

and

$$N_x = e^{xy} + xye^{xy},$$

which implies that the differential equation is exact.

To obtain F we compute $F = \int M \, dx$ and $F = \int N \, dy$.

$$F = \int M \, dx = \int y e^{xy} \, dx = e^{xy} + h_1(y)$$

where $h_1(y)$ is an unknown function of y. Similarly,

$$F = \int N \, dy = \int x e^{xy} + \sin y \, dy = e^{xy} - \cos y + h_2(x)$$

where $h_2(x)$ is an unknown function of x.

For F to equal both simultaneously, we must have $h_2(x) = 0$ and $h_1(y) = -\cos y$.

Thus $F(x, y) = e^{xy} - \cos y$ and hence,

$$e^{xy} - \cos y = C$$

is an implicit solution to the DE for any C.

To solve the initial value problem, when $x = 0$ we must have $y = \pi$ or $e^0 - \cos \pi = C$ which implies that $C = 2$. Thus,

$$e^{xy} - \cos y = 2$$

solves the initial value problem. $\qquad \square$

Exercises

Use the Exactness Test to Determine if the DE is exact.

1. $y^2\, dx + x\, dy = 0$

2. $(x^2 + y^2)\, dx + (2xy + \cos y)\, dy = 0$

3. $s\, dr + r\, ds = 0$

4. $\arctan(y)\, dx + \dfrac{x}{1 + y^2}\, dy = 0$

Use the Exactness Test to show the DE is exact, then solve it.

5. $(\sqrt{y} + 2x \tan y)\, dx + \left(\dfrac{x}{2\sqrt{y}} + x^2 \sec^2 y \right) dy = 0$

6. $(2xy^4 - y^3 + \cos(2x))\, dx + (4x^2 y^3 - 3y^2 x - 2y)\, dy = 0$

7. $\left(\dfrac{y}{x} - 3y^2 + x^3 \right) dx + (\ln x - 6xy)\, dy = 0$

8. $\left(\sqrt{x^2 + y^2} + \dfrac{x^2}{\sqrt{x^2 + y^2}} \right) dx + \dfrac{xy}{\sqrt{x^2 + y^2}} dy = 0$ (Hint: one integration is easier, use the easy one to backward engineer the harder one)

9. $(\cos(xy) - xy \sin(xy))\, dx + (-x^2 \sin(xy) + y)\, dy = 0$ (Hint: one integration is easier, use the easy one to backward engineer the harder one))

Use the Exactness Test to show the DE is exact, then solve the initial value problem.

10. $2xy^3\, dx + 3x^2 y^2\, dy = 0, \quad y(1) = 2$

11. $\left(y \cos(xy) - y - 1 \right) dx + \left(x \cos(xy) - x - y \right) dy = 0, \quad y\!\left(\dfrac{1}{2}\right) = \pi$

12. $(y^2 - 2xe^y)\, dx + (2xy - x^2 e^y)\, dy = 0, \quad y(2) = 0$

13. (a) Show that $xy^4 dx + 4x^2 y^3 \, dy = 0$ is not exact.

(b) Multiply the DE by $\frac{1}{x}$ and show that the resulting DE is exact.

(c) Solve the DE from (b). Does the solution in (b) solve the original DE (in (a))?

2.4   Applications-Exponential Growth, Cooling

Many things grow (decay) as fast as exponential functions. In general, if a quantity grows or decays at a rate proportional to quantity itself, then it will exhibit exponential behavior.

Exponential Growth/Decay

Consider a quantity Q which is known to change at a rate proportional to itself:

$$\frac{dQ}{dt} = rQ$$

(r is the constant of proportionality) Then

$$Q(t) = Ke^{rt}, \quad \text{with K} = Q(0) \tag{2.11}$$

Proof: Clearly if $Q(t) = Ke^{rt}$, then

$$Q'(t) = rKe^{rt} = rQ(t).$$

Plugging $t = 0$ into $Q(t) = Ke^{rt}$, we see that $Q(0) = K$.   □

Note in equation (2.11) that r is determined by the rate at which the quantity grows or decays. Clearly this depends upon the units with which Q is measured and with which time is measured.

Radioactive Decay

It is well-known that radioactive materials decay at a rate proportional to the amount of material present.

Example 2.9 *(Carbon Dating)* *Carbon-14 has a half-life of 5,730 years. A fossil is found to have 10% of its original Carbon-14. Determine the age of the fossil.*

Solution: Let $Q(t)$ be the amount of Carbon-14 present in the fossil at time t, where $t = 0$ is the time of the fossilized animal. The half-life is clearly related to the quantity r, since both concern the rate of decay. So we first determine r by noticing that when $t = 5730$ there will be one-half of the original amount of Carbon-14, or

$$Q(5730) = \frac{1}{2}Q(0).$$

In particular

$$Q(0)e^{5730r} = \frac{1}{2}Q(0),$$

so

$$e^{5730r} = \frac{1}{2},$$

$$r = \frac{\ln(\frac{1}{2})}{5730} = -\frac{\ln 2}{5730}.$$

Next, we solve for the value of t that yields 10% of its original Carbon-14. That is we wish to solve

$$Q(t) = .1Q(0)$$

for t.

$$Q(0)e^{-\frac{\ln 2}{5730}t} = \frac{1}{10}Q(0),$$

$$e^{-\frac{\ln 2}{5730}t} = \frac{1}{10},$$

$$-\frac{\ln 2}{5730}t = \ln(\frac{1}{10}),$$

$$t = \frac{5730\ln 10}{\ln 2} \approx 19034.648 \text{ years } \square$$

Population Models

A biological population that is not subject to resource limitations will grow at a rate proportional to itself. This is entirely believable, since if the size of the population increases by a factor of k then we would expect the growth rate to also increase by a factor of k (roughly, the number of babies doubles if the population doubles). Hence, if $P(t)$ the population of a specific species at time t, then it is reasonable that $\frac{dP}{dt} = rP$. This is called a Malthusian or exponential population model. Such populations grow exponentially, and this model works reasonably well when the population in question is not close to being constrained by a lack of resources. Clearly, if $r > 0$ the population is growing and if $r < 0$ the population is dying.

Example 2.10 *(Bacteria Counts) Escherichia coli (E. Coli) is measured in colony forming units per milliliter (CFU/mL). In 'ideal' circumstances E. Coli in dairy milk has a doubling time of roughly 20 minutes. Pasteurized milk is*

Grade A if it has less than 1,000 CFU/mL. How long will it take for a gallon of milk with 1,000 CFU/mL to reach 1,000,000 CFU/mL (which is considered harmful) when left in 'ideal' circumstances?

Solution: Let $Q(t)$ be the CFU/mL of E. Coli at time t (minutes). The doubling rate determines r (just as half-life determines r).

In particular

$$Q(20) = 2Q(0)$$

so

$$Q(0)e^{20r} = 2Q(0)$$

or

$$r = \frac{1}{20}\ln 2.$$

We wish to solve for t in

$$Q(t) = Q(0)e^{rt} = 10^6,$$

where $Q(0) = 10^3$. Solving, we obtain

$$10^3 e^{\frac{\ln 2}{20}t} = 10^6$$

so

$$t = 20 \cdot \frac{\ln 10^3}{\ln 2} \approx 199.32 \text{ minutes}$$

or about 3 hours and 19.32 minutes. $\square$

Financial Models

Money invested at interest generally grows in proportion to the amount of money that is invested (principal).

Example 2.11 (Saving for Retirement) *A 25 year-old has inherited $ 50,000.00 and plans to invest it in an investment that pays 5% annual interest for 40 years (meaning after one year, she will earn exactly 5% of $ 50,000.00). How much will be in the bank after 40 years?*

Solution: Let $Q(t)$ be the dollar value of the investment at time t (years). Then $\frac{dQ}{dt} = rQ$ and $Q(0) = 50,000$.

It is somewhat clear that the interest rate will dictate r, but how? After 1 year, we will have added exactly 5% of $Q(0)$, so

$$Q(1) = Q(0) + (.05)Q(0)$$

or

$$Q(0)e^r = 1.05Q(0).$$

So,

$$r = \ln 1.05 \approx 0.04879016417.$$

Note this is not equal to .05. The reason is that our model assumes that interest grows continuously, whereas in reality it is only added once a year.

So, to solve the problem, we want

$$Q(40) = 50000e^{40 \ln 1.05} = 50000 \cdot (1.05)^{40} = \$351,999.44 \ \square$$

The previous example shows that one should always realize a mathematical model does not represent the actual quantities precisely. In truth, bacteria and investments do not grow *continuously*. Moreover, the actual quantities involved in population models can only be integers and the amount of dollars is at best measured to two decimal places. However, these models do an excellent job reflecting the real quantities that they are modeling.

Calibrating Parameters in a Model

The constants that appear in a model are also called *parameters*. Usually, the parameters in a model must be fit from the given data, much as the constant C is determined from initial data. The next example shows how this is done.

Example 2.12 *(Saving for Retirement)* *A 25 year-old has inherited* $\$50,000.00$ *and plans to invest it in an investment that pays 5% annual interest for 40 years (meaning after one year, she will earn exactly 5% of $ 50,000.00). Additionally, she will deposit $4,000.00 per year into the account at the end of each year.*
   (a) Find the correct value of D in the model $\frac{dQ}{dt} = rQ + D$
   (b) How much will be in the bank after 40 years?
   (c) How much of this is interest?

Solution: (a) Let $Q(t)$ be the dollar value of the investment at time t (years). Then $\frac{dQ}{dt} = rQ + D$ and $Q(0) = 50,000$. As in the previous example,

$$r = \ln 1.05 \approx 0.04879016417.$$

Solving the linear DE $\frac{dQ}{dt} - rQ = D$, we obtain:

$$Q(t) = \frac{\int e^{-rt} D \ dt + C}{e^{-rt}} = \frac{-\frac{D}{r} e^{-rt} + C}{e^{-rt}} = Ce^{rt} - \frac{D}{r},$$

where $r = \ln 1.05$.

As usual $Q(0)$ will help us solve for C.

$$50000 = C - \frac{D}{\ln(1.05)}$$

so $C = 50000 + \frac{D}{\ln(1.05)}$.

Since we know that after one year, there should be an additional \$2,500 (interest) along with a deposit of \$4000 (additional deposit), we have $Q(1) = 56500$.

So we can now solve for D.

$$56500 = \left(50000 + \frac{D}{\ln(1.05)} \right) e^{\ln(1.05)1} - \frac{D}{\ln(1.05)}$$

So

$$56500 - 50000(1.05) = D \left(\frac{1.05}{\ln(1.05)} - \frac{1}{\ln(1.05)} \right)$$

$$D = \frac{56500 - 52500}{\left(\frac{.05}{\ln(1.05)} \right)}$$

$$D = \frac{4000}{\left(\frac{.05}{\ln(1.05)} \right)}$$

$$D = 80000 \ln(1.05)$$

$$D \approx 3903.213133$$

(note that D is NOT 4000!).

(b)

$$Q(40) = (50000 + 80000) e^{\ln(1.05)40} - 80000 \approx \$835,198.53$$

(c) Since she deposited \$ 4000 forty times and had \$50,000 originally, she contributed a total of $50000 + 40 \times 4000 = 50000 + 160000 = \$210,000$. So the

difference is interest, which is $ 625,198.53! That is over half of a million dollars!
□

Newton's Law of Cooling

Newton's Law of Cooling (see Example 1.6) states the rate of change of the temperature of an object is proportional to the difference of the temperature T of that object and the ambient temperature M of the room. If $T(t)$ denotes the temperature of the object at time t, then this translates to

$$\frac{dT}{dt} = k(T - M),$$

where k is a constant that determined by the cooling rate. This is a first order linear ODE, which can easily be solved.

Example 2.13 *(Forensics) A corpse is found at noon in a room that is held at a constant temperature of* $70°F$. *The body is determined to have a temperature of* $78°F$. *One half hour later, the body cools to* $76°F$. *Assuming that the deceased person had a body temperature of* $98.6°F$ *at the time of death, determine the time of death to the nearest minute.*

Solution: Let $T(t)$ denotes the temperature of the object at time t (in hours), where $t = 0$ is noon. By Newton's Law of Cooling

$$\frac{dT}{dt} = k(T - 70)$$

or

$$\frac{dT}{dt} - kT = -70k.$$

Using the first order linear formula:

$$T(t) = \frac{\int -70ke^{-kt} \, dt + C}{e^{-kt}}$$

$$T(t) = (70e^{-kt} + C)e^{kt} = 70 + Ce^{kt}$$

We recover C by using the information $T(0) = 78$, which gives $C = 8$
Next, we recover K by using the information $T(\frac{1}{2}) = 76$

$$76 = 70 + 8e^{\frac{1}{2}k},$$

So

$$k = 2\ln\left(\frac{6}{8}\right) = \ln\left(\frac{9}{16}\right).$$

Finally, we solve for $T(t) = 98.6$ for t (expecting a negative value of t).

$$98.6 = 70 + 8e^{\ln\left(\frac{9}{16}\right)t} = 70 + 8\left(\frac{9}{16}\right)^t$$

So

$$\frac{28.6}{8} = \left(\frac{9}{16}\right)^t$$

$$t = \frac{\ln(28.6) - \ln 8}{\ln 9 - \ln 16} \approx -2.214189 \text{ hours}$$

So the person died approximately 2 hours and 13 minutes ago which would have been about $9:47$ AM. $\quad\square$

Newton's Law of Cooling works equally well in a heating situation, where the initial temperature of the object is below the ambient temperature.

Exercises

Use the fact that C^{14} has a half life of around 5730 years.

1. How old is a fossil with 25% of its original C^{14}?

2. Suppose that a researcher is confident that a fossil has somewhere between $15 - 35\%$ of its original C^{14}. What is the range of possible ages of this fossil?

3. The production of the first atomic bomb also produced the byproduct of 1881 Ci (Curie) of the radioactive isotope Radium-226 which has a half life of 1600 years. This material is currently stored in Lewiston, NY (my hometown!!). Even 1 Curie of this material is extremely hazardous, but compute how long it will take for the 1881 Ci of Ra^{228} to decay to 1 Curie.

4. A rabbit population doubles every 6 months. If the colony starts with 500 rabbits, how long will it take to reach 1500, assuming exponential growth?

5. (a) A \$5000 investment is made for 30 years at 8% annual interest. How much will the investment be worth? (Hint: r is not .08 and must be calibrated using $Q(1) = 1.08Q(0)$).

   (b) Additionally, the investor plans to add \$1200 at the end of each year to his investment. How much will the investor have after 40 years? [Hint: use the (linear) DE $\frac{dQ}{dt} = K + rQ$, where r is solved for in part (a) and K is determined $Q(1) = \$6600$.]

6. A loan of \$100,000 is taken out at 5% annual interest (continuous interest) for 30 years.

   (a) Assume it is paid off at a continuous (and constant) rate K. Determine K so that the loan is completely paid off in 30 years. [Hint: use the (linear) DE $\frac{dQ}{dt} = rQ - K$, where r is 0.05, and K is determined $Q(30) = 0$.]

   (b) What monthly payment does this value of K correspond to?

7. Mr. Boddy's body is discovered in a walk-in freezer that is held at $34°F$. His body has a temperature of $42°F$. An hour later, his body cools to $38°F$. Find how long ago Mr. Boddy died from the initial time that he was discovered. (Hint: Make $t = 0$ the time the body is $42°F$.)

   [Assume that living persons have a body temperature of $98.6°F$.]

8. An indoor tank is filled with $52°F$ water. The ambient temperature of the room is $72°F$.

   (a) If it takes one hour for the temperature to rise one degree from $52°F$ to $53°F$, how long will it take for the temperature of the water to reach $71°F$?

   (b) Will the temperature of the water ever reach $72°F$? Explain.

9. How long will it take a $40°F$ glass of milk to heat to an undrinkable $60°F$ in a room that is held at $72°F$? (Assume that after 1 minute, the milk increases from $40°F$ to $43°F$).

10. A room's temperature is changing throughout the day and is given by $M(t) = 70 + 3\sin(\frac{t}{2})$ where t is measured in minutes. A pot of boiling water is brought into this room to cool (at $t = 0$). After 1 minute, it cools from $212°F$ to $192°F$. Find its temperature for any t. Plot the temperature of the water and the temperature of the room against time over 2 hours.

2.5   More Applications (Mixing, Logistic Model)

Mixture Problems

Differential equations also lend themselves to mixing problems as in the following example. We provide a few tips for writing the necessary differential equation.

Tips for Setting Up a Model Using a Differential Equation

1. Define the unknown function so that it solves the problem at hand, in almost all examples, quantities should be absolute quantities (avoid relative quantities ,e.g. percentages). Clearly and specifically define the units of both dependent and independent variables.

2. Write a differential equation based on the rates at which the desired quantity is changing (a sketch may help). In general, avoid using initial data information, as this will be used for the initial condition. Recall that the rate of change of the quantity in question can often be thought of as

$$\text{rate of increase } - \text{ rate of decrease}$$

3. Write the initial condition.

Example 2.14 *(A Mixture Problem)* *A 200 liter vat initially contains a 10% solution of Hydrogen Peroxide and 90 % water. If a 1 % hydrogen peroxide-water solution is added to the vat at a constant rate of 2L/min and the mixture is drained off at the same rate, determine how long it will take until the mixture reaches a concentration of 3% Hydrogen Peroxide. The vat is well-stirred, so assume that the Hydrogen Peroxide is uniformly distributed in the vat.*

Solution: Let $M(t)$ denote the liters of hydrogen peroxide in the tank at time t, where t is measured in minutes.

Since the mixture entering the vat at a rate of 2L/min is 1% hydrogen peroxide, the rate of hydrogen peroxide is entering the tank is .02L/min.

The rate of mixture exiting the tank is also 2L/min. However, not all of this is hydrogen peroxide. With the assumption that the mixture is uniformly mixed, for any time t, we know that the amount of hydrogen peroxide in 1L of the mixture in the vat is $\dfrac{M(t)}{200}$. Thus

$$\frac{dM}{dt} = rate\ of\ increase\ of\ M -\ rate\ of\ decrease\ of\ M = .02 - 2\frac{1}{200}M.$$

The initial condition is $M(0) = 20$.

The DE is a first order linear

$$\frac{dM}{dt} + \frac{1}{100}M = \frac{2}{100}$$

Solving:

$$M(t) = \frac{\int \frac{2}{100}e^{\frac{t}{100}}\ dt + C}{e^{\frac{t}{100}}}$$

$$= \left(2e^{\frac{t}{100}} + C\right) \cdot e^{-\frac{t}{100}} = 2 + Ce^{-\frac{t}{100}}$$

As usual, the initial condition gives the particular value of C, namely $M(0) = 20$ implies that $C = 18$, so

$$M(t) = 2 + 18e^{-\frac{t}{100}}$$

We want to know at what t will the solution reach a 3% concentration of hydrogen peroxide. Hence, we seek to solve

$$\frac{M(t)}{200} = .03$$

or

$$M(t) = 6.$$

So we are solving

$$6 = 2 + 18e^{-\frac{t}{100}}$$

for t.

We obtain

$$4 = 18e^{-\frac{t}{100}}$$

or

$$\frac{4}{18} = e^{-\frac{t}{100}},$$

so

$$t = -100\ln\left(\frac{2}{9}\right) \approx 150.40774\ \text{minutes}$$

or about $2\frac{1}{2}$ hours. □

We solve another mixture type problem.

Example 2.15 *(Another Mixture Problem)* *A 50 gallon storage tank initially holds a saltwater solution containing 100 grams of salt dissolved in 50 gallons of water. A salt/water mixture is pumped in at a rate of 2 gallon per minute. The incoming mixture has 4 grams of salt per gallon of mixture. The salt water is also is pumped out at a rate of 3 gallons per minute. How much salt is in the tank when the tank is half full? What is the concentration of salt (grams/gallon) in the mixture that is exiting the tank as it becomes empty?*

Solution: Let $M(t)$ denote the amount of sale (in grams) in the tank at time t, where t is measured in minutes.

Since the mixture entering the tank at a rate of 2 gallons/min and each gallon contains 4 gram of salt, there are 8 grams of salt entering per minute.

The rate of mixture exiting the tank is 3 gallons/min. Assuming that the salt is evenly dispersed in the tank, the amount of salt in one gallon of mixture in the tank is given by $\dfrac{M(t)}{50 - t}$ where $50 - t$ is the total amount (in gallons) of solution in the tank. So in all, the rate of salt exiting the tank is $\dfrac{3M}{50 - t}$

Thus

$$\frac{dM}{dt} = rate\ of\ increase\ of\ M - rate\ of\ decrease\ of\ M = 8 - \frac{3}{50 - t}M.$$

Or

$$\frac{dM}{dt} + \frac{3}{50 - t}M = 8.$$

Solving

$$M(t) = \frac{\int 8e^{-3\ln(50 - t)}\ dt + C}{e^{-3\ln(50 - t)}}$$

$$M(t) = \frac{\int \frac{8}{(50 - t)^3}\ dt + C}{\frac{1}{(50 - t)^3}}$$

$$M(t) = \left(\frac{4}{(50 - t)^2} + C\right)(50 - t)^3 = 4(50 - t) + C(50 - t)^3$$

As usual, the initial condition gives the particular value of C, namely $M(0) = 100$ implies that $100 = 200 + 50^3 C$

$$C = \frac{-100}{50^3}$$

so

$$M(t) = 4(50 - t) - \frac{100}{50^3}(50 - t)^3$$

Since the tank will be half full when $t = 25$ we have

$$M(25) = 4(25) - \frac{100}{50^3}25^3 = 100 - 100(\frac{1}{2})^3 = 100 + 12.5 = 87.5 \text{ grams}$$

The number of grams of salt per gallon at any time t is given by $\dfrac{M(t)}{50 - t} = 4 - \dfrac{100}{50^3}(50 - t)^2$, so as the tank empties, $t = 50$ so the the solution has a concentration of 4 grams of salt per gallon. $\square$

Logistic Model

We have already investigated population models where the population is not limited by resources. As in Equation (2.11), the population behaves as an exponential function which is unbounded. In this section, we seek to create a model that takes resource limitations into account. We start with a concrete example and then present the general logistic model.

Example 2.16 *(A rabbit colony) A colony of 1000 rabbits lives in a field that can support 5000 rabbits. Write a differential equation that takes the resource limitation into account.*

Solution: Let $P(t)$ be the population at time t. As described previously, in the absence of resource limitations, a good model is:

$$\frac{dP}{dt} = rP.$$

We want

$$\frac{dP}{dt} > 0 \quad \text{when} \quad 0 < P < 5000$$

and

$$\frac{dP}{dt} < 0 \quad \text{when} \quad P > 5000$$

This implies that we want

$$\frac{dP}{dt} = 0 \quad \text{when} \quad P = 5000.$$

A simple way to achieve this is allow r in Equation (2.11) to be a function of P, and make r negative if $P > 5000$. We do this by letting $r = k(5000 - P)$, and we obtain the DE

$$\frac{dP}{dt} = k(5000 - P)P \tag{2.12}$$

where k is a constant that is determined by the growth rate of the population. □

In the previous example, the population was limited by resources to 5000 rabbits. This value is called a carrying capacity or limiting population. In general,

---

The Logistic Differential Equation

A population P at time t with a carrying capacity of P_∞ is modeled by the **logistic differential equation** (or logistic growth model)

$$\frac{dP}{dt} = kP(P_\infty - P) \tag{2.13}$$

where $k > 0$ is a constant that is determined by the growth rate of the population.

We solve this separable differential equation (2.13) as follows:

$$\frac{dP}{P(P_\infty - P)} = k\, dt$$

(with $P(t) \neq 0$ and $P(t) \neq P_\infty$, which are two constant solutions).

$$\int \frac{dP}{P(P_\infty - P)} = \int k\, dt$$

Using partial fractions on the left integrand:

$$\int \frac{\frac{1}{P_\infty}}{P} + \frac{\frac{1}{P_\infty}}{P_\infty - P} \, dP = \int k \, dt$$

$$\frac{1}{P_\infty} \int \frac{1}{P} + \frac{1}{P_\infty - P} \, dP = \int k \, dt$$

$$\frac{1}{P_\infty} \left(\ln|P| - \ln|P_\infty - P| \right) = kt + C$$

$$\ln|P| - \ln|P_\infty - P| = P_\infty(kt + C)$$

$$\ln\left| \frac{P}{P_\infty - P} \right| = P_\infty(kt + C)$$

$$\left| \frac{P}{P_\infty - P} \right| = e^{P_\infty(kt+C)}$$

Note that we can let $\widetilde{C} = e^{P_\infty C}$

$$\left| \frac{P}{P_\infty - P} \right| = \widetilde{C} e^{P_\infty kt}$$

where $\widetilde{C} > 0$. Dropping absolute value on the left amounts to dropping the positivity condition on $\widetilde{C}$ so

$$\frac{P}{P_\infty - P} = \widetilde{C} e^{P_\infty kt}$$

or

$$P = (P_\infty - P)\, \widetilde{C} e^{P_\infty kt}$$

$$P + P\widetilde{C} e^{P_\infty kt} = P_\infty \widetilde{C} e^{P_\infty kt}$$

$$P = \frac{P_\infty \widetilde{C} e^{P_\infty kt}}{1 + \widetilde{C} e^{P_\infty kt}}$$

Multiplying by

$$\frac{e^{-P_\infty kt}}{e^{-P_\infty kt}}$$

we obtain

$$P(t) = \frac{P_\infty \widetilde{C}}{\widetilde{C} + e^{-P_\infty kt}}$$

We can solve for $\widetilde{C}$ in terms of $P(0)$, the initial population,

$$P(0) = \frac{P_\infty \widetilde{C}}{\widetilde{C} + 1}$$

So

$$P(0)(\widetilde{C} + 1) = P_\infty \widetilde{C}$$

giving

$$\widetilde{C} = \frac{P(0)}{P_\infty - P(0)}$$

or

$$P(t) = \frac{P_\infty \frac{P(0)}{P_\infty - P(0)}}{\frac{P(0)}{P_\infty - P(0)} + e^{-P_\infty kt}}$$

which simplifies to

$$P(t) = \frac{P_\infty P(0)}{P(0) + \left[P_\infty - P(0)\right] e^{-P_\infty kt}}$$

Note that the constant solution $P(t) = 0$ is obtained if $P(0) = 0$, and the constant solution $P(t) = P_\infty$ is obtained if $P(0) = P_\infty$.

We have proven that

Solutions of the Logistic Differential Equation

The solutions to

$$\frac{dP}{dt} = kP(P_\infty - P)$$

are given by

$$P(t) = \frac{P_\infty P_0}{P_0 + \left[P_\infty - P_0\right] e^{-P_\infty kt}} \qquad (2.14)$$

where P_0 is $P(0)$ and P_∞ is the limiting population (carrying capacity).

One should note that as t approaches infinity, $P(t)$ limits to the limiting population P_∞ or

$$\lim_{t \to \infty} P(t) = \lim_{t \to \infty} \frac{P_\infty P_0}{P_0 + \left[P_\infty - P_0\right] e^{-P_\infty kt}} = P_\infty.$$

Example 2.17 *(A rabbit colony-continued) A colony of 1000 rabbits lives in a field that can support 5000 rabbits. After 1 month, the colony reaches 1050 rabbits. Use a logistic model to predict when the population will have 2500 rabbits.*

Solution: Let $P(t)$ be the population of rabbits at time t (in months). By Equation (2.14)

$$P(t) = \frac{P_\infty P_0}{P_0 + [P_\infty - P_0] e^{-P_\infty k t}}$$

where $P_0 = 1000$ and $P_\infty = 5000$. So

$$P(t) = \frac{1000 \cdot 5000}{1000 + 4000 e^{-5000kt}} = \frac{5000}{1 + 4e^{-5000kt}}$$

We need to compute k using the information that $P(1) = 1050$. So

$$1050 = P(1) = \frac{5000}{1 + 4e^{-5000k}}$$

which solving for k gives

$$k = \frac{-1}{5000} \ln\left(\frac{1}{4}\left(\frac{5000}{1050} - 1\right)\right) = \frac{-1}{5000} \ln\left(\frac{79}{84}\right)$$

So

$$P(t) = \frac{5000}{1 + 4e^{\ln\left[\frac{79}{84}\right]t}}$$

$$P(t) = \frac{5000}{1 + 4\left[\frac{79}{84}\right]^t}$$

We solve for $P(t) = 2500$ and obtain

$$2500 = \frac{5000}{1 + 4\left[\frac{79}{84}\right]^t}$$

$$1 + 4\left[\frac{79}{84}\right]^t = 2$$

$$4\left[\frac{79}{84}\right]^t = 1$$

$$\left[\frac{79}{84}\right]^t = \frac{1}{4}$$

$$t = \ln\left(\frac{1}{4}\right)\frac{1}{\ln\left[\frac{79}{84}\right]} \approx 22.589 \ \text{ months}\square$$

A Note: in the previous problem, had we not cancelled (especially inside the exponential), we would have introduced serious round off error in the computation and our calculation would have been far off

Exercises

1. A 100 gallon tank holds 50 gallons of water with 75 grams of salt dissolved into it. If a mixture with 6 grams per gallon of salt is pumped in at 2 gallons per hour and the mixture is drained at 1 gallon per hour, determine how much salt will be in the tank when it is full.

2. Suppose a 5,000 gallon tank contains a 1,000 gallons of a juice mixture which is 90% juice and 10 % water. Pure water is pumped in at 5 gallons per second and the mixture is drained 4 gallons per second. How long will it take until the percentage of juice in the tank is 50%?

3. A tank starts with 500 liters of water with 1 kg of salt dissolved in it. A salt and water mixture with concentration 0.1kg/L is poured into the tank at a rate of 2L/min. The mixture is drained at 4L/min. Assuming that the mixture is well stirred is (uniformly distributed)

   (a) Write a DE together with an initial condition that describes the amount of salt $m(t)$ in the tank at time t.

   (b) Solve the DE/IVP.

4. A large mixing container initially holds 100 gallons of liquid, 98% water and 2 % alcohol. Into the container is pumped a mixture containing 5 % alcohol at a rate of 6 gallons per minute. The mixture is constantly drained off at a rate of 1 gallon per minute. Write an ODE/IVP that describes the rate of change of the amount of alcohol (in gallons) in the tank for any time t and solve the DE/IVP to get the amount of alcohol in the tank for any time t.

5. The world's (human) population went from 4 billion in 1975 to six billion in 2000. If the carrying capacity is 20 billion, estimate at what year the human population will reach 19 billion (and will start to be limited by resources). [Hint: measure P in billions, i.e. $P(0) = 4$].

6. A certain ant colony has grown from 5000 to 6000 ants in 6 months. The colony has a carrying capacity of 10000. Find how long it will take until the colony reaches a size of 9000.

7. Prove (using calculus) that the change of concavity of all solutions of Equation (2.14) with $0 < P_0 < P_\infty$ take place when $P = \frac{1}{2}P_\infty$. Explain how this can be used to estimate a population's carrying capacity given only a graph of the population over time.

8. An acidic solution flows at a constant rate of 6 liters per minute into a large tank that initially held 50 liters of acid solution in which was dissolved 5 kg of pure acid. The solution inside the tank is kept well stirred and flows out of the tank at the same rate. If the concentration of acid in the solution entering the tank is 0.5 kg/L, determine the mass of acid in the tank after t minutes.

2.6   Slope Fields and Euler's Method

Any first order differential equation can be thought of geometrically. In particular the DE

$$\frac{dy}{dx} = f(x, y)$$

specifies a specific slope for any point (x, y) given by the function f.

A **slope field** is simply a plot of the slope

$$\frac{dy}{dx} = f(x, y)$$

for several points (x, y) represented by a vector having slope $f(x, y)$ based at (x, y), usually plotted with a fixed change in x (or so that all vectors have a uniform length).

Example 2.18 *For $\frac{dy}{dx} = \frac{1}{2}(x^2 - y)$, plot the associated vectors at the points: $(1, 2), (1, 3), (2, 2), (2, 3)$, where the vectors are pictured having a change in x of 1 ($\Delta x = 1$).*

Solution:

(x, y)	$\frac{dy}{dx}$
$(1, 2)$	$-\frac{1}{2}$
$(1, 3)$	-1
$(2, 2)$	1
$(2, 3)$	$\frac{1}{2}$

The vector at the point $(1, 2)$, is to have slope $-\frac{1}{2}$, and it is to have $\Delta x = 1$. Since $\Delta y = \frac{dy}{dx} \Delta x$ the vector will be drawn from $(1, 2)$ to $(1 + \Delta x, 2 + \Delta y)$ so it will have terminal point at $(1 + 1, 2 - 0.5) = (2, 1.5)$. Plotting all four vectors, we obtain Figure (2.1).

□

In general, one would plot many points to get a better idea of the behavior of the differential equation, however it is also clear that one cannot plot too many points or the resulting plot will be a cluttered mess. In Figure 2.2, we show a slope field plot for $\frac{dy}{dx} = \frac{1}{2}(x^2 - y)$ where all vectors are plotted with the same length.

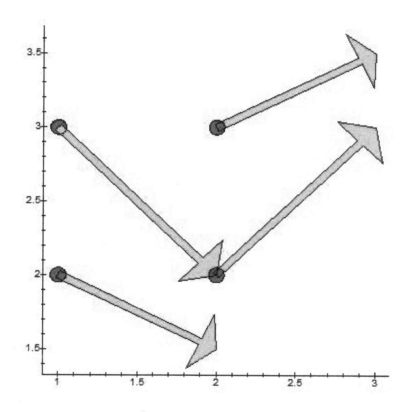

Figure 2.1: A slope field plot for four points of $\frac{dy}{dx} = \frac{1}{2}(x^2 - y)$.

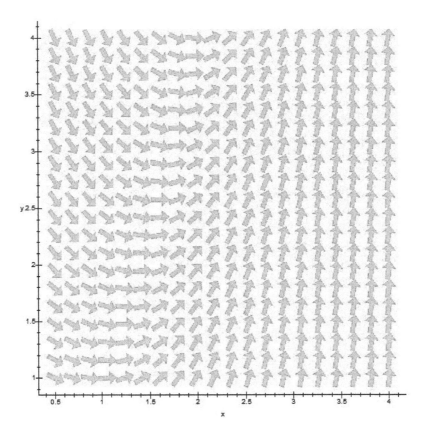

Figure 2.2: A slope field plot of $\frac{dy}{dx} = \frac{1}{2}(x^2 - y)$.

Figure 2.2 gives us a good idea how the solutions of this differential equation will behave. Geometrically, the graph of a solution to a differential equation is always tangent to vectors on the slope field. (Informally, if the slope field were like the wind, the graph of a solution would be the path a particle would take when blown by the wind).

We can use this idea to propose a method to approximate solutions to differential equations, namely at the point (x_0, y_0) we draw a line segment with slope $f(x_0, y_0) = \frac{dy}{dx}$. If we increase x_0 by Δx, i.e. let $x_1 = x_0 + \Delta x$ then we can estimate that an approximate solution will have $y_1 = y_0 + \Delta x f(x_0, y_0)$ (see figure). This is the idea that drives the numerical method called Euler's method (pronounced 'Oiler').

Euler's Method to Approximate Solutions of First Order ODE

Consider the differential equation

$$\frac{dy}{dx} = f(x, y)$$

with initial condition $y(x_0) = y_0$ (i.e., we want the graph of the solution to pass through (x_0, y_0)).

Let $x_1 = x_0 + \Delta x$.

$$y_1 = y_0 + f(x_0, y_0)\Delta x$$

Similarly, let $x_2 = x_1 + \Delta x$.

$$y_2 = y_1 + f(x_1, y_1)\Delta x$$

and, in general $x_{j+1} = x_j + \Delta x$ and

$$y_{j+1} = y_j + f(x_j, y_j)\Delta x \tag{2.15}$$

.

Then the sequence of points $(x_0, y_0), (x_1, y_1), (x_2, y_2), ..., (x_n, y_n)$ is an approximation to the graph of the solution with initial condition $y(x_0) = y_0$.

In Euler's Method, the quantity Δx is referred to as the **step size**.

Example 2.19 *Use Euler's method to approximate the solution to*

$$\frac{dy}{dx} = y - y^2 = y(1 - y)$$

with initial condition $y(0) = 2$. Use step size of $\Delta x = 0.3$ and estimate $y(3)$.

Solution: With a step size of $\Delta x = 0.3$, it will take 10 steps to be able to approximate $y(3)$. Note that $f(x, y) = y - y^2$. Starting with $x_0 = 0$ and $y_0 = 2$ we generate $x_1 = x_0 + \Delta x = 1 + 0.3 = 1.3$ and

$$y_1 = y_0 + f(x_0, y_0)\Delta x = 2 + f(0, 2)(0.3) = 2 + (2 - 4)(.3) = 2 - 0.6 = 1.4$$

Continuing, $x_2 = x_1 + \Delta x = 1.3 + 0.3 = 1.6$ and

$$y_2 = y_1 + f(x_1, y_1)\Delta x = 1.4 + f(1.3, 1.4)(0.3) = 1.4 + (1.4 - 1.4^2)(0.3) = 1.232$$

We produce the remainder of the values below:

$step$	x	y	$\frac{dy}{dx}$	Δx
0	0	2	-2	0.3
1	0.3	1.4	-0.56	0.3
2	0.6	1.232	-0.285824	0.3
3	0.9	1.1462528	-0.167642682	0.3
4	1.2	1.095959995	-0.105168316	0.3
5	1.5	1.0644095	-0.068558084	0.3
6	1.8	1.043842075	-0.045764203	0.3
7	2.1	1.030112814	-0.031019596	0.3
8	2.4	1.020806935	-0.021239864	0.3
9	2.7	1.014434976	-0.014643345	0.3
10	3.0	1.010041973	-0.010142814	0.3

So $y(3) \approx 1.010041973$.   □

This process can easily implemented on a spreadsheet or into computer code. In Figure (2.3), we plot the sequence of points $(x_0, y_0), (x_1, y_1), ..., (x_{10}, y_{10})$ and the corresponding slope field plot which illustrates how Euler's method works.

We note that the differential equation in Example (2.19) could have been solved explicitly (see the previous section).

Example 2.20 *Use Euler's method to approximate the solution to*

$$\frac{dy}{dx} = \frac{1}{10}y(x^2 + y)$$

with initial condition $y(2) = 1$. Use step size of $\Delta x = 0.05$ and estimate $y(3)$.

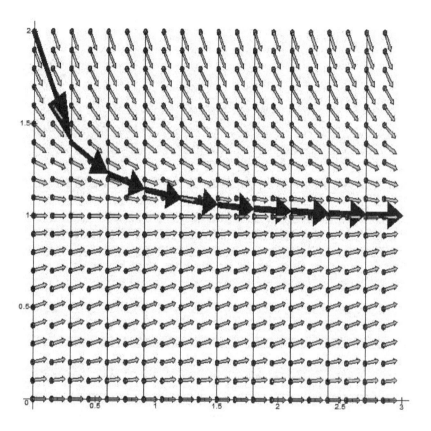

Figure 2.3: A slope field plot of $\frac{dy}{dx} = y - y^2$ together with a plot thick black arrows that with initial and terminal points given by the points generated in Example (2.19).

Solution: With a step size of $\Delta x = 0.05$, it will take 20 steps to be able to approximate $y(3)$. Note that $f(x, y) = \frac{1}{10} y(x^2 + y)$. Starting with $x_0 = 2$ and $y_0 = 1$ we generate $x_1 = 2 + 0.05 = 2.05$ and

$$y_1 = y_0 + f(x_0, y_0)\Delta x = 1 + f(1, 2)(0.05) = 1 + \frac{1}{10}5(.05) = 1.025$$

Continuing, $x_2 = x_1 + \Delta x = 2.1$ and

$$y_2 = y_1 + f(x_1, y_1)\Delta x = 1.025 + \frac{1}{10}1.025(2.05^2 + 1.025)(.05) = 1.051790$$

We produce the remainder of the values below:

step	x	y	$\frac{dy}{dx}$	Δx
0	2	1	.5	.05
1	2.05	1.025	.535818	.05
2	2.1	1.051790	.574466	.05
3	2.15	1.080514	.616218	.05
4	2.2	1.111325	.661385	.05
5	2.25	1.144394	.710313	.05
6	2.3	1.179910	.763391	.05
7	2.35	1.218079	.821056	.05
8	2.4	1.259132	.883801	.05
9	2.45	1.303322	.952184	.05
10	2.5	1.350931	1.026834	.05
11	2.55	1.402273	1.108465	.05
12	2.6	1.457696	1.197891	.05
13	2.65	1.517591	1.296036	.05
14	2.7	1.582393	1.403961	.05
15	2.75	1.652591	1.522877	.05
16	2.8	1.728735	1.654180	.05
17	2.85	1.811444	1.799478	.05
18	2.9	1.901418	1.960631	.05
19	2.95	1.999449	2.139801	.05
20	3	2.106439	2.339504	.05

So $y(3) \approx 2.106439$.   □

Figure 2.4 shows the plot of this approximation.

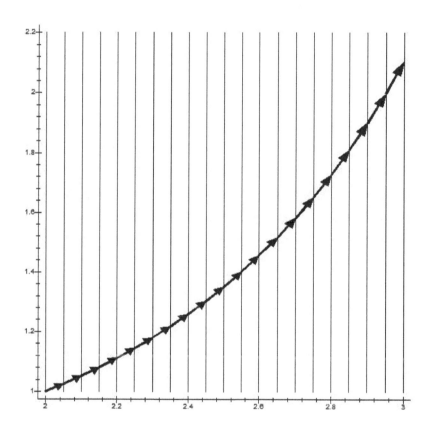

Figure 2.4: The plots of the points generated in Example (2.20).

One should realize that there is a bit of an art-form in choosing Δx. Clearly one wants Δx to be small enough so that $y(x_0 + \Delta x) \approx y(x_0) + \frac{dy}{dx}\Delta x$, however if one chooses Δx too small then one can accumulate round off error and can actually get less precise approximations (or run into excessive computation time).

Exercises

1. Draw the slope field vectors for the DE

$$\frac{dy}{dx} = \frac{xy + 1}{x + 2y}$$

   at the points $(1, 1), (1, 2), (2, 2), (2, 1)$. Make each vector have a change in x of 1.

2. Use Euler's method to approximate the solution to the IVP $\frac{dy}{dx} = \sqrt{x + 2y}$ where $y(0) = 1$ using 10 steps with $\Delta x = 0.2$ to estimate $y(2)$. Could you have obtained the precise solution for this IVP using the methods we have learned?

3. Use Euler's method to approximate the solution to the IVP $\frac{dy}{dx} = 3y(1 - y)$ where $y(0) = 2$ using 4 steps with $\Delta x = 1$ to estimate $y(2)$. Explain what went wrong (if anything).

4. Draw a slope field for

$$\frac{dy}{dx} = x^2 - y^3$$

   .

   (a) Can you predict the long term behavior of solutions?

   (b) Use Euler's method to approximate the solutions with initial conditions $y(0) = 0$, $y(0) = 1$, and $y(0) = 2$ with at least 100 steps (use technology) to compute where these solutions will be for $t = 2$.

5. Consider

$$\frac{dy}{dx} = \frac{1}{2}y - x^2, \ y(0) = 1$$

   (a) Use a table with 4 steps to estimate $y(2)$

   (b) Use a spreadsheet or computer to use 200 steps to estimate $y(2)$

   (c) Use a spreadsheet or computer to use 2000 steps to estimate $y(2)$

   (d) Are the answers close?

   (e) You could have also solved this DE explicitly. Compare the actual value of $y(2)$ with the estimates in (a), (b), and (c).

6. Draw a direction field for (use a graph with $0 \leq y \leq 6$)

$$\frac{dy}{dx} = y(y-2)(4-y)$$

Use your field to predict the behavior of a solution with IC's when $x \to \infty$

(a) $y(0) = 1$

(b) $y(0) = 2$

(c) $y(0) = 3$

(d) $y(0) = 4$

(e) $y(0) = 5$

7. Consider

$$\frac{dy}{dx} = \frac{1}{2}y - x^2, \ y(0) = 1$$

(a) Use a table with 4 steps to estimate $y(2)$

(b) Use a spreadsheet or computer to use 200 steps to estimate $y(2)$

(c) Use a spreadsheet or computer to use 2000 steps to estimate $y(2)$

(d) Solve this DE explicitly. Compare the actual value of $y(2)$ with the estimates in (a), (b), and (c).

Chapter 3

Second Order ODE

3.1 Introduction

In this chapter, we study second order ODE. Usually in a calculus or physics class one studies the second order equation:

$$\frac{d^2y}{dt^2} = -32\frac{ft}{sec^2}$$

where $y(t)$ is the height (in feet) of a free falling object at time t. This second order ODE can be solved simply by integrating twice to obtain:

$$y(t) = -16t^2 + C_1t + C_2$$

where C_1 and C_2 are constants that are determined by initial conditions (such as initial height and initial velocity. In particular, it is typical that general solutions to a second order ODE will have two free constants.

In this chapter we will study second order DE that cannot simply be solved by integrating.

3.2 Homogeneous with Constant Coefficients

The first type of second order DE that we will study are homogeneous second order linear ODE with constant coefficients which are differential equations that can be written in the form

$$a\frac{d^2y}{dx^2} + b\frac{dy}{dx} + cy = 0$$

for constants a, b, c which can be written as

$$ay'' + by' + cy = 0$$

where the independent variable is unspecified.

Let us consider the differential equation

$$y'' + 2y' + 3y = 0 \tag{3.1}$$

It is quite clear that any polynomial will **not** solve this equation, since if y is a degree n polynomial then y'' and y' will be polynomials with degree strictly less than n. Therefore, the degree n term that appears in the $3y$ in the above ODE will never cancel with any terms from y'' or $2y'$, hence we will never obtain zero.

Thus, a reasonable expectation is to search for functions whose derivatives look reasonably similar to the original function in hopes of cancellations. This leads us to consider the class of exponential functions $y = e^{rt}$ for for a fixed constant r.

Plugging into the left-hand side of the differential equation (3.1), we obtain

$$r^2 e^{rt} + 2r e^{rt} + 3 e^{rt}$$

which factors as

$$e^{rt}(r^2 + 2r + 3).$$

Since $e^{rt} > 0$ for all t, if we intend for this quantity to be equal to zero (for all t) we must have

$$r^2 + 2r + 3 = 0.$$

Solving for r, we obtain $r = -3$ and $r = 1$. In other words we have shown that $y_1 = e^{-3t}$ and $y_2 = e^t$ both solve differential equation (3.1).

Since y_1 and y_2 both solve the DE (3.1), one might guess that

$$y = c_1 y_1 + c_2 y_2$$

will also solve the DE, where c_1 and c_2 are constant. This is indeed the case since:

$$y' = c_1 y_1' + c_2 y_2'$$

and

$$y'' = c_1 y_1'' + c_2 y_2''.$$

Plugging into the left-hand side of the DE (and factoring) we obtain:

$$y'' + 2y' + 3y =$$

$$c_1(y_1'' + 2y_1' + 3y_1) + c_2(y_2'' + 2y_2' + 3y_2)$$

However, the items in the parentheses above are zero since both y_1 and y_2 solve DE (3.1).

Real Distinct Roots Case

What we have done in the above discussion will work in general. In particular:

Homogeneous constant coefficients–real distinct roots case

If the homogeneous second order linear DE with constant coefficients

$$ay'' + by' + cy = 0 \tag{3.2}$$

has associated polynomial

$$ar^2 + br + c = 0$$

and suppose that this polynomial has two distinct real roots r_1 and r_2, then the general solution of (3.2) is given by

$$y = c_1 e^{r_1 t} + c_2 e^{r_2 t}$$

Note: The associated polynomial to DE (3.2) is called the characteristic polynomial of (3.2).

Example 3.1 *Solve the DE*

$$y'' - 5y' + 6y = 0$$

Solution: We form the characteristic polynomial.

$$r^2 - 5r + 6 = (r - 2)(r - 3)$$

which clearly has roots $r_1 = 2$ and $r_2 = 3$. Hence the general solution is

$$y = c_1 e^{2t} + c_2 e^{3t} \quad \square$$

Example 3.2 *Solve the DE*

$$\frac{d^2y}{dx^2} - 4\frac{dy}{dx} - y = 0$$

Solution: We form the characteristic polynomial.

$$r^2 - 4r - 1.$$

This polynomial does not factor easily so we use the quadratic formula:

$$r_{1,2} = \frac{-b \pm \sqrt{b^2 - 4ac}}{2a} = \frac{4 \pm \sqrt{20}}{2} = 2 \pm \sqrt{5}$$

so $r_1 = 2 + \sqrt{5}$ and $r_2 = 2 - \sqrt{5}$. Thus the general solution is

$$y = c_1 e^{(2+\sqrt{5})x} + c_2 e^{(2-\sqrt{5})x}.$$

Note that the independent variable was specified to be x from the DE.   □

Recall from algebra (or by staring at the quadratic formula) that a quadratic polynomial

$$ar^2 + br + c$$

has two distinct real roots if, and only if,

$$b^2 - 4ac > 0.$$

We note that this method even works if one of the roots of the quadratic formula is itself zero as the next example shows:

Example 3.3 *Solve the DE/IVP*

$$\frac{d^2z}{dt^2} - 3\frac{dz}{dt} = 0, \quad z(0) = 1, \quad z'(0) = 2$$

Solution: We form the characteristic polynomial.

$$r^2 - 3r = r(r - 3)$$

which clearly has roots $r_1 = 0$ and $r_2 = 3$. Thus the general solution is

$$z = c_1 e^{0t} + c_2 e^{3t}$$

or

$$z = c_1 + c_2 e^{3t}.$$

Next, we use the initial conditions to find the correct values of the constants:

$$1 = z(0) = c_1 + c_2$$

and $z' = 3c_2 e^{3t}$ so $z'(0) = 2$ implies that $c_2 = \frac{2}{3}$ and hence (by $c_1 + c_2 = 1$) we see that $c_1 = \frac{1}{3}$. So the particular solution to the IVP is

$$z = \frac{1}{3} + \frac{2}{3}e^{3t}. \quad \square$$

Repeated Roots Case

As mentioned at the beginning of the chapter, one would expect a second order DE to have a general solution with two free constants. If equation

$$ay'' + by' + cy = 0$$

has a characteristic polynomial

$$ar^2 + br + c$$

that has two repeated real roots (which happens exactly when $b^2 - 4ac = 0$), then our method previous runs into problems.

For instance, if we consider

$$y'' + 6y' + 9y = 0$$

we obtain

$$r^2 + 6r + 9 = (r + 3)^2$$

which has repeated roots $r_1 = -3$ and $r_2 = -3$.

When we form the general solution we get

$$y = c_1 e^{-3t} + c_2 e^{-3t}$$

but this can be written as

$$y = (c_1 + c_2)e^{-3t} = K_1 e^{-3t}.$$

So the trouble here is that we actually only have one free constant (so we are missing solutions to this DE). It turns out that another solution to this DE is obtained by $z = te^{-3t}$ (one could check by differentiating and using the product rule). In a later section, we will motivate where this other solution comes from, but one might arrive at it from judicious guessing.

In the general repeated roots case:

Homogeneous constant coefficients–repeated roots case

If the homogeneous second order linear DE with constant coefficients

$$ay'' + by' + cy = 0 \qquad (3.3)$$

has associated polynomial

$$ar^2 + br + c = 0$$

and suppose that this polynomial has two repeated real roots $r_1 = r_2$ (which occurs exactly when $b^2 - 4ac = 0$). then the general solution of (3.3) is given by

$$y = c_1 e^{r_1 t} + c_2 t e^{r_1 t}$$

Example 3.4 *Solve the DE*

$$y'' + 10y' + 25y = 0$$

Solution: We form the characteristic polynomial.

$$r^2 + 10r + 25 = (r + 5)^2$$

which clearly has repeated roots $r_1 = -5$ and $r_2 = -5$. Hence the general solution is

$$y = c_1 e^{-5t} + c_2 t e^{-5t} \quad \square$$

Complex Roots Case

Our last case is if the DE

$$ay'' + by' + cy = 0$$

has a characteristic polynomial

$$ar^2 + br + c$$

that has complex roots (which happens exactly when $b^2 - 4ac < 0$). One should recall from precalculus (or by staring long enough at the quadratic formula) that complex roots come in conjugate pairs, which means if $\alpha + \beta i$ is a complex root, then so is $\alpha - \beta i$.

In particular, if $b^2 - 4ac < 0$ then the complex roots are:

$$r_{1,2} = -\frac{b}{2a} \pm \frac{\sqrt{4ac - b^2}}{2a} i$$

(this is because $\sqrt{b^2 - 4ac} = \sqrt{-1(4ac - b^2)} = \sqrt{4ac - b^2}\, i$)

If we label $r_{1,2} = \alpha \pm \beta i$ then we can obtain the general solution to the DE as described below:

Homogeneous constant coefficients–repeated roots case

If the homogeneous second order linear DE with constant coefficients

$$ay'' + by' + cy = 0 \tag{3.4}$$

has associated polynomial

$$ar^2 + br + c = 0$$

and suppose that this polynomial has complex roots $r_{1,2} = \alpha \pm \beta i$ (which occurs exactly when $b^2 - 4ac < 0$). then the general solution of (3.4) is given by

$$y = c_1 e^{\alpha t} \cos(\beta t) + c_2 e^{\alpha t} \sin(\beta t)$$

or

$$y = c_1 e^{-\left(\frac{b}{2a}\right)t} \cos\left(\frac{\sqrt{4ac - b^2}}{2a} t\right) + c_2 e^{-\left(\frac{b}{2a}\right)t} \sin\left(\frac{\sqrt{4ac - b^2}}{2a} t\right)$$

Example 3.5 *Solve the DE*

$$y'' + 2y' + 2y = 0$$

Solution: We form the characteristic polynomial.

$$r^2 + 2r + 2$$

which, by the quadratic formula has roots

$$r_{1,2} = \frac{-b \pm \sqrt{b^2 - 4ac}}{2a} = \frac{-2 \pm \sqrt{4 - 4(2)}}{2} = \frac{-2 \pm \sqrt{-4}}{2}$$

$$= \frac{-2 \pm \sqrt{4}i}{2} = -1 \pm i$$

so $r_{1,2} = 1 \pm i$ so $\alpha = -1$ and $\beta = 1$.

Hence the general solution is

$$y = c_1 e^{-t} \cos t + c_2 e^{-t} \sin t \quad \square$$

Example 3.6 *Solve the DE*

$$y'' + 4y = 0$$

Solution: We form the characteristic polynomial.

$$r^2 + 4$$

which, by the quadratic formula has roots $r_{1,2} = \pm 2\,i$ so $\alpha = 0$ and $\beta = 2$.

Hence the general solution is

$$y = c_1 \cos(2t) + c_2 \sin(2t) \quad \square$$

Exercises

Find general solutions for each of the DEs, note that the roots are real

1. $y'' + 4y' + 4y = 0$

2. $y'' + 3y' + 2y = 0$

3. $y'' + 6y' - 7y = 0$

4. $z'' + 4z' + z = 0$

5. $z'' - z = 0$

6. $z'' + 2z' = 0$

Find general solutions for each of the DEs

7. $y'' + 4y' + 6y = 0$

8. $y'' + 6y' + 10y = 0$

9. $y'' - 7y = 0$

10. $y'' + 7y = 0$

11. $z'' + z' + z = 0$

12. $z'' + 2z' + 5z = 0$

Find particular solutions for each of the DE/IVP

13. $y'' + 4y' + 3y = 0$, $y(0) = 1$, $y'(0) = -1$

14. $y'' + 3y' + 2y = 0$, $y(0) = 0$, $y'(0) = 1$

15. $y'' + 8y' + 17y = 0$, $y(0) = 0$, $y'(0) = 1$

16. $y'' - 9y' + 20y = 0$, $y(0) = 4$, $y'(0) = -2$

17. $y'' + 16y = 0$, $y(0) = 4$, $y'(0) = -2$

18. $z'' - 6z = 0$, $z(0) = 1$, $z'(0) = 2$

19. $z'' - 6z = 0$, $z(0) = 0$, $z'(0) = 0$

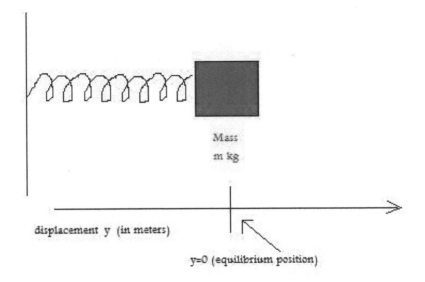

Figure 3.1: A spring mass system

3.3   Application-Spring Mass Systems

Second order differential equations arise naturally when the second derivative of a quantity is known. For example, in many applications the acceleration of an object is known by some physical laws like Newton's Second Law of Motion $F = ma$. One particularly nice application of second order differential equations with constant coefficients is the model of a spring mass system.

Suppose that a mass of m kg is attached to a spring. From physics, Hooke's Law states that if a spring is displaced a distance of y from its equilibrium position, then the force exerted by the spring is a constant $k > 0$ multiplied by the displacement of the y. In other words,

$$F_{spring} = -ky.$$

The negative sign above is due to the fact that the force will always be in the opposite direction of the displacement.

Undamped Springs (no friction)

We are now in a position to formulate a model of a spring/mass system. By Newton's Second Law,

$$F = ma$$

and we realize that $a = y''(t)$. So we obtain the second order differential equation

$$my'' = -ky$$

which we rewrite as

$$my'' + ky = 0$$

where $m > 0$ and $k > 0$.

Of course we can solve this system for all values of m, k since it is a homogeneous linear second order DE with constant coefficients.

It has general solution:

$$y(t) = c_1 \cos(\sqrt{\frac{k}{m}}t) + c_2 \sin(\sqrt{\frac{k}{m}}t)$$

The long term behavior of this spring/mass system as suggested from the general solution above is that the mass will oscillate forever, which is not realistic. This suggests that our model is missing some key physical feature. Indeed, we have neglected frictional forces. However, if it were possible to have no friction, then the model reflects what we would expect.

Example 3.7 *A spring with spring constant 18N/m is attached to a 2kg mass with negligible friction. Determine the period that the spring mass system will oscillate for any non-zero initial conditions.*

Solution: From above, we have a spring mass system modelled by the DE

$$2y'' + 18y = 0$$

which has general solution given by

$$y(t) = c_1 \cos(\sqrt{\frac{18}{2}}t) + c_2 \sin(\sqrt{\frac{18}{2}}t) = c_1 \cos(3t) + c_2 \sin(3t)$$

Since period of $\cos t$ is 2π, then the period of $\cos(3t)$ and $\sin(3t)$ is $\frac{2\pi}{3}$. Therefore, the period of $c_1 \cos(3t) + c_2 \sin(3t)$ is also $\frac{2\pi}{3}$. $\square$

Note: The frequency of $\cos(\beta t)$ is often defined two (different) ways, one way is $frequency = \frac{1}{period} = \frac{\beta}{2\pi}$. Another similar definition is the angular frequency of $\cos(\beta t)$ which is simply β. We suggest avoiding frequencies altogether and working with the period, since, then, there is no confusion.

Converting $c_1 \cos(\beta t) + c_2 \sin(\beta t)$ into phasor form $A\cos(\beta t - \phi)$
In this section we show how to convert

$$y = c_1 \cos(\beta t) + c_2 \sin(\beta t) \qquad (3.5)$$

into the form $y = A\cos(\beta t - \phi)$.

Using the difference formulas from trigonometry

$$A\cos(X - \phi) = A\cos\phi\cos X + A\sin\phi\sin X$$

and taking $X = \beta t$ in this formula, and matching with formula (3.6) we obtain, $c_1 = A\cos\phi$ and $c_2 = A\sin\phi$.

Note that

$$c_1^2 + c_2^2 = A^2(\cos^2\phi + \sin^2\phi) = A^2$$

so

$$A = \sqrt{c_1^2 + c_2^2}.$$

So long $c_1 \neq 0$, we have

$$\frac{c_2}{c_1} = \frac{A\sin\phi}{A\cos\phi} = \tan\phi$$

and so note that $\phi = \arctan(\frac{c_2}{c_1})$ if $c_1 > 0$ and $\phi = \arctan(\frac{c_2}{c_1}) + \pi$ if $c_1 < 0$.

You may realize that the values of A and ϕ are simply the *polar coordinates* (r, θ) of the point (c_1, c_2).

In more compact form, so long as $c_1 \neq 0$

$$\phi = \arctan(\frac{c_2}{c_1}) + \frac{\pi}{2}(1 - \frac{c_1}{|c_1|})$$

Note that if $c_1 = 0$ then (from polar coordinates) we see that $A = |c_2|$ and $\phi = \frac{\pi}{2}$ if $c_2 > 0$ and $\phi = -\frac{\pi}{2}$ if $c_2 < 0$.

We summarize below:

Phasor Form

To convert

$$y = c_1 \cos(\beta t) + c_2 \sin(\beta t) \tag{3.6}$$

into the form $y = A \cos(\beta t - \phi)$:

$$A = \sqrt{c_1^2 + c_2^2}$$

and

$$\phi = \begin{cases} \arctan\left(\frac{c_2}{c_1}\right) & c_1 > 0 \\ \arctan\left(\frac{c_2}{c_1}\right) + \pi & c_1 < 0 \\ \frac{\pi}{2} & c_1 = 0, \ c_2 > 0 \\ -\frac{\pi}{2} & c_1 = 0, \ c_2 < 0 \end{cases}$$

In other words, (A, ϕ) are the polar coordinates of the rectangular point (c_1, c_2).

Note: In the above formula, c_1 must always be the coefficient of the sine term and c_2 must be the coefficient of the cosine term. Also, the sine and cosine functions must have the same argument.

Example 3.8 *Convert each of the following to phase angle (phasor) notation.*
(a)

$$y = 4\cos(3t) - 4\sin(3t)$$

(b)

$$y = -\cos(2t) + \sqrt{3}\sin(2t)$$

(c)

$$y = -7\sin(t)$$

Solution:
(a) We see that $c_1 = 4$ and $c_2 = -4$ so

$$A = \sqrt{c_1^2 + c_2^2} = \sqrt{4^2 + 4^2} = \sqrt{16 + 16} = \sqrt{32} = 4\sqrt{2}$$

$$\phi = \arctan\left(\frac{-4}{4}\right) = \arctan(-1) + 0 = -\frac{\pi}{4}$$

So
$$y = 4\sqrt{2}\cos(3t + \frac{\pi}{4})$$

.

(b) We see that $c_1 = -1$ and $c_2 = \sqrt{3}$ so

$$A = \sqrt{c_1^2 + c_2^2} = \sqrt{1^2 + (\sqrt{3})^2} = \sqrt{1 + 3} = \sqrt{4} = 2$$

$$\phi = \arctan\left(\frac{\sqrt{3}}{-1}\right) = \arctan(-\sqrt{3}) + \pi = -\frac{\pi}{3} + \pi = \frac{2\pi}{3}$$

So
$$y = 2\cos(2t - \frac{2\pi}{3})$$

(c) We see that $c_1 = 0$ and $c_2 = -7$. So we cannot use the formula involving arc tangent. But when we plot the point $(0, -7)$ we see that in polar it is $r = 7$ and $\theta = -\frac{\pi}{2}$ so $A = 7$ and $\phi = -\frac{\pi}{2}$ and

$$y = 7\cos(t + \frac{\pi}{2}) \quad \square$$

Example 3.9 *A spring with spring constant 4N/m is attached to a 1kg mass with negligible friction. If the mass is initially displaced to the right of equilibrium by 0.5m and has an initial velocity of 1 m/s toward equilibrium. Compute the amplitude of the oscillation.*

Solution:

As before, the spring mass system corresponds to the DE

$$y'' + 4y = 0.$$

Since the mass is displaced to the right of equilibrium by $0.5m$, we have $y(0) = \frac{1}{2}$. Since the mass an initial velocity of 1 m/s toward equilibrium (to the left) $y'(0) = -1$.

Solving the spring mass system, we obtain the general solution

$$y(t) = c_1\cos(2t) + c_2\sin(2t).$$

$y(0) = \frac{1}{2}$ gives $c_1 = \frac{1}{2}$ and since

$$y'(t) = -2c_1\sin(2t) + 2c_2\cos(2t),$$

we see that $y'(0) = -1$ implies that $-1 = 2c_2$ or $c_2 = -\frac{1}{2}$.

So

$$y(t) = \frac{1}{2}\cos(2t) - \frac{1}{2}\sin(2t).$$

Converting to phase/angle notation, we see $A = \sqrt{(\frac{1}{2})^2 + (-\frac{1}{2})^2} = \frac{\sqrt{2}}{2}$ so the amplitude of oscillation will be $\frac{\sqrt{2}}{2}m$.

Note that $\phi = -\frac{\pi}{4}$ since the polar angle of $(\frac{1}{2}, -\frac{1}{2})$ is $-\frac{\pi}{4}$. □

Exercises

Convert to phase/angle (phasor) form

1. $y = 2\cos t - 2\sin t$

2. $y = -1\cos(6t) - \sqrt{3}\sin(6t)$

3. $y = \cos(4t)$

4. $y = -12\sin(\sqrt{2}t)$

5. A spring with spring constant $2N/m$ is attached to a 1kg mass with negligible friction. Compute the period of the oscillation for any non-zero initial conditions.

6. A spring with spring constant $16N/m$ is attached to a 1kg mass with negligible friction. If the mass is initially displaced to the left of equilibrium by $0.25m$ and has an initial velocity of 1 m/s toward equilibrium. Compute the amplitude and period of the oscillation.

7. A spring with spring constant $16N/m$ is attached to a 1kg mass with negligible friction. If the mass is initially at equilibrium with an initial velocity of 2 m/s toward the left. Compute the amplitude and period of the oscillation.

8. A spring with spring constant $2N/m$ is attached to a 1kg mass with negligible friction. If the mass is initially 1m to the left of equilibrium with no initial velocity. Compute the amplitude and period of the oscillation.

3.4 Unforced Spring Mass Systems with Friction

In this section we introduce friction onto the spring mass system from the last section. To model friction, we realize that the force of friction always opposes the direction of motion. In other words, if an object is moving to the right, it will experience a frictional force to the left. Moreover, we will work under the assumption that the force due to friction is proportional to the velocity (that is if the velocity doubles, then so will the force due to friction, etc.)

Using these assumptions

$$F_{friction} = -by'(t)$$

where $y'(t)$ is the velocity of the mass at time t and b is the constant of proportionality called the friction constant. (Note that when y is in meters, t is seconds and mass is kilograms, the constant b is measured in Newton·sec/meter.) Combining the frictional force and the force from Hooke's Law, we obtain

$$F_{total} = -by' - ky$$

and by Newton's Law of motion

$$my'' = -by' - ky$$

or

$$my'' + by' + ky = 0$$

where $m > 0$ is mass, $b > 0$ is the friction constant, $k > 0$ is the spring. (Note that the units of force need not be Newtons and y need not be meters, but units must be compatible, for example k will be units of force per units of distance, etc.). We obtain a second order linear homogeneous ODE with constant coefficients. Also, as before, there are three cases to consider based on the roots of the characteristic polynomial $mr^2 + br + k = 0$ which lead to three different types of spring mass systems.

Overdamped-Real Distinct Roots

First we analyze the case where the spring mass system has characteristic polynomial $mr^2 + br + k = 0$ that has real distinct roots, namely when $b^2 - 4mk > 0$.

By the quadratic formula, these roots are:

$$r_1 = \frac{-b + \sqrt{b^2 - 4mk}}{2m}$$

$$r_2 = \frac{-b - \sqrt{b^2 - 4mk}}{2m}$$

and the general solution will be given by

$$y = c_1 e^{r_1 t} + c_2 e^{r_2 t}$$

Since $b, m, k > 0$ it is clear that $r_2 < 0$.

Also, since

$$4mk > 0$$

we see that

$$0 > -4mk$$

so adding b^2 to both sides

$$b^2 > b^2 - 4mk$$

and taking square roots of both sides,

$$b > \sqrt{b^2 - 4mk}$$

subtracting b from both sides,

$$0 > -b + \sqrt{b^2 - 4mk}$$

which implies r_1 is itself negative since $r_1 = \frac{-b + \sqrt{b^2 - 4mk}}{2m}$ and $m > 0$.

Thus, we see that no matter what the initial conditions are, we can see that

$$\lim_{t \to \infty} y(t) = \lim_{t \to \infty} c_1 e^{r_1 t} + c_2 e^{r_2 t} = 0.$$

This makes physical sense to us, since if there is friction, the spring will always limit to the equilibrium position.

Also, any non zero solution $y(t)$ to an overdamped spring mass problem can have at most one time t_* where $y(t_*) = 0$.

To see this, suppose $y(t_*) = 0$, then

$$0 = c_1 e^{r_1 t_*} + c_2 e^{r_2 t_*}$$

Solving for t_* we see that

$$c_1 e^{r_1 t_*} = -c_2 e^{r_2 t_*}$$

$$e^{r_1 t_*} = -\frac{c_2}{c_1} e^{r_2 t_*}$$

$$e^{r_1 t_*} e^{-r_2 t_*} = -\frac{c_2}{c_1}$$

$$e^{(r_1 - r_2) t_*} = -\frac{c_2}{c_1}$$

$$(r_1 - r_2) t_* = \ln(-\frac{c_2}{c_1})$$

$$t_* = \frac{1}{r_1 - r_2} \ln(-\frac{c_2}{c_1}). \tag{3.7}$$

Thus, the only possible time for $y(t) = 0$ is given by t_* above. Note that this value may not even exist if the argument of the logarithm is negative, which would imply that $y(t) \neq 0$ for all t.

A similar argument shows that any non-zero solution $y(t)$ to an overdamped spring mass system can have at most one time t where $y'(t) = 0$ (which implies that y can have at most one local maximum or minimum) and at most one time t where $y''(t) = 0$ (which implies that $y(t)$ can have at most one inflection point).

Critically Damped-Real Repeated Roots

Next, we analyze the case where the spring mass system has characteristic polynomial $mr^2 + br + k = 0$ that has real repeated roots, namely when $b^2 - 4mk = 0$.

This implies that the roots are $r_{1,2} = -\frac{b}{2m}$ and that the general solution to the homogeneous spring mass system is given by

$$y(t) = c_1 e^{-\frac{b}{2m}t} + c_2 t e^{-\frac{b}{2m}t}$$

Notice that

$$\lim_{t \to \infty} y(t) = \lim_{t \to \infty} c_1 e^{-\frac{b}{2m}t} + c_2 t e^{-\frac{b}{2m}t} = 0$$

by applying L'Hôpital's Rule on the term $c_2 t e^{-\frac{b}{2m}t}$. Again this makes physical sense, since friction will cause the mass to limit to equilibrium.

As in the overdamped case, if $y(t_*) = 0$ then

$$y(t_*) = c_1 e^{-\frac{b}{2m}t_*} + c_2 t_* e^{-\frac{b}{2m}t_*} = (c_1 + c_2 t_*) e^{-\frac{b}{2m}t_*}$$

so
$$t_* = -\frac{c_1}{c_2}.$$

In other words, a non-zero solution to a critically damped spring mass system can pass through the equilibrium position at most once (or never if $c_2 = 0$). Similar arguments show that (as in the overdamped case) that a non-zero solution $y(t)$ to a critically damped spring mass system can have at most one time t where $y'(t) = 0$ (which implies that y can have at most one local maximum or minimum) and at most one time t where $y''(t) = 0$ (which implies that $y(t)$ can have at most one inflection point).

Note that a spring mass system that is critically damped is not physically a possibility since it is unlikely that $b^2 - 4mk$ exactly equals zero.

Overdamped/Critically Damped Spring Mass Systems

The spring mass system
$$my'' + by' + ky = 0 \tag{3.8}$$

is called overdamped if
$$b^2 - 4mk > 0$$

and critically damped if
$$b^2 - 4mk = 0.$$

All non-zero solutions to overdamped or critically damped spring mass systems:

- limit to equilibrium as $t \to \infty$,

- pass through the equilibrium position at most once (possible not at all),

- have at most one maxima (or none at all),

- have at most one point of inflection (or none at all).

- limit to $\pm\infty$ as $t \to -\infty$,

In light of the above result, the plots of solutions of critically damped or overdamped systems tend to look similar to the samples that are plotted below:

Example 3.10 *A spring with spring constant 4N/m is attached to a 1kg mass with friction constant 5Ns/m. If the mass is initially displaced to the right of equilibrium by 0.1m and has an initial velocity of 1 m/s toward equilibrium.*

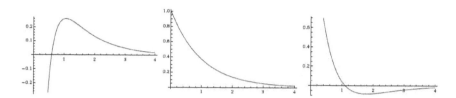

Figure 3.2: Plots of three different solutions of overdamped/critically damped systems

(a) Determine if the mass passes through the equilibrium position, if so determine when it does so.

(b) Determine if the displacement has any local extrema for t > 0.

Solution:

We see that $m = 1$, $b = 5$ and $k = 4$. Also, we see that $b^2 - 4mk = 5^2 - 4 \cdot 1 \cdot 4 = 25 - 16 = 9$, so the spring mass system is overdamped.

The roots of the characteristic polynomial (which is $r^2 + 5r + 4 = 0$) are $r = 1$ and $r = 4$. So the general solution is

$$y(t) = c_1 e^{-t} + c_2 e^{-4t}$$

Plugging in the initial conditions:

$$\frac{1}{10} = y(0) = c_1 + c_2$$

and

$$-1 = y'(0) = -c_1 - 4c_2$$

(Note that since the mass is initially located to the right of equilibrium and is moving toward equilibrium (left), $y'(0)$ is negative).

Adding we obtain,

$$-\frac{9}{10} = -3c_2$$

so

$$c_2 = \frac{3}{10}$$

which implies that

$$c_1 = -\frac{2}{10}.$$

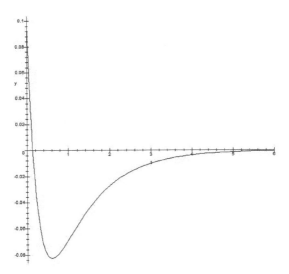

Figure 3.3: Plot of the solution to Example (3.10)

So the particular solution is

$$y(t) = -\frac{2}{10}e^{-t} + \frac{3}{10}e^{-4t}.$$

Now, we can solve for the time t_* when the mass is at equilibrium:

$$0 = -\frac{2}{10}e^{-t_*} + \frac{3}{10}e^{-4t_*}.$$

$$2e^{3t_*} = 3$$

so the answer to (a) is

$$t_* = \frac{1}{3}\ln(\frac{3}{2}) \approx 0.135155036$$

We can resolve (b) by taking the derivative of our particular solution

$$y'(t) = \frac{2}{10}e^{-t} - \frac{12}{10}e^{-4t}$$

and setting it equal to zero (the derivative always exists, so our only possible critical values are ones where the derivative is zero). Solving $y'(t) = 0$, we get

$$2e^{3t} = 12$$

$$t = \frac{1}{3}\ln 6 \approx 0.597253156,$$

we could see that at this time, our function has an absolute minimum from either the second derivative test, or from knowing that the solution must limit to zero and can have at most one extrema. At any rate, it is at this time that the mass is displaced the furthest the left (most negative). □

Underdamped-Complex Roots

Next, we analyze the case where the spring mass system has characteristic polynomial $mr^2 + br + k = 0$ that has complex roots, namely when $b^2 - 4mk < 0$.

This implies that the roots of the characteristic polynomial are complex $r_{1,2} = \alpha \pm \beta i = -\frac{b}{2m} \pm \frac{\sqrt{4mk-b^2}}{2m} i$ and that the general solution to the homogeneous spring mass system is given by

$$y(t) = c_1 e^{-\frac{b}{2m}t}\cos(\frac{\sqrt{4mk-b^2}}{2m}t) + c_2 e^{-\frac{b}{2m}t}\sin(\frac{\sqrt{4mk-b^2}}{2m}),$$

which can be rewritten in phase angle notation as

$$y(t) = Ae^{-\frac{b}{2m}t}\cos(\frac{\sqrt{4mk-b^2}}{2m}t - \phi)$$

where $A = \sqrt{c_1^2 + c_2^2}$ and $\tan\phi = \frac{c_2}{c_1}$. The value $\beta = \frac{\sqrt{4mk-b^2}}{2m}$ is called the *pseudo frequency*, since the function is a periodic function multiplied by a decaying exponential function.

In any case, we see for any fixed initial conditions, using the squeeze theorem from calculus, since

$$-Ae^{-\frac{b}{2m}t} \le y(t) \le Ae^{-\frac{b}{2m}t} \tag{3.9}$$

then

$$\lim_{t\to\infty} -Ae^{-\frac{b}{2m}t} \ge \lim_{t\to\infty} y(t) \le \lim_{t\to\infty} Ae^{-\frac{b}{2m}t}$$

so

$$0 \le \lim_{t\to\infty} y(t) \le 0$$

and

$$\lim_{t\to\infty} y(t) = 0.$$

Moreover, equation (3.9) allows us to see that $y(t)$ will oscillate between the two curves $-Ae^{-\frac{b}{2m}t}$ and $Ae^{-\frac{b}{2m}t}$. Unlike the overdamped and critically damped cases, the mass will pass through the equilibrium position an infinite number of times (since there are an infinite number of t values that solve $\cos(\frac{\sqrt{4mk-b^2}}{2m}t-\phi) = 0$ or $\frac{\sqrt{4mk-b^2}}{2m}t - \phi = n\pi + \frac{\pi}{2}$ for n an integer).

Underdamped Spring Mass Systems

The spring mass system

$$my'' + by' + ky = 0 \tag{3.10}$$

is called under damped if

$$b^2 - 4mk < 0$$

All non-zero solutions to underdamped spring mass systems:

- limit to equilibrium as $t \to \infty$,

- pass through the equilibrium position infinitely may times,

- have infinitely many maxima,

The figure below shows a solution to an underdamped spring mass DE and the bounding functions.

Example 3.11 *A spring with spring constant $18N/m$ is attached to a 2kg mass with friction constant $4Ns/m$. If the mass has initially position 1 meter to the right of equilibrium and has no initial velocity:*
(a) Find the solution,
(b) Express the solution in phase/angle form,
(c) Plot the solution together with its two bounding curves.

Solution: From above, we see that $m = 2$, $b = 4$ and $k = 20$. Also, we see that $b^2 - 4mk = 16 - 4 \cdot 1 \cdot 20 = 16 - 80 = -64$, so the spring mass system is underdamped.

The roots of the characteristic polynomial (which is $r^2+2r+10 = (r+1)^2+9$) are $r_{1,2} = -1 \pm 3i$. So the general solution is

$$y(t) = c_1 e^{-t} \cos(3t) + c_2 e^{-t} \sin(3t).$$

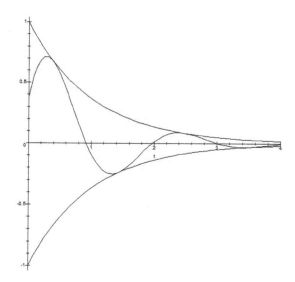

Figure 3.4: Plot of a solution to an underdamped spring mass system with bounds

Using the initial conditions, $c_1 = 1$ and

$$y'(t) = -c_1 e^{-t} \cos(3t) - 3c_1 e^{-t} \sin(3t) - c_2 e^{-t} \sin(3t) + 3c_2 e^{-t} \cos(3t).$$

or

$$0 = -c_1 + 3c_2$$

or

$$c_2 = \frac{1}{3}$$

Thus

$$y(t) = e^{-t} \cos(3t) + \frac{1}{3} e^{-t} \sin(3t),$$

which in phase/angle form is

$$y(t) = \sqrt{1 + \frac{1}{9}} e^{-t} \cos(3t - \arctan(\frac{1}{3}))$$

$$= \frac{\sqrt{10}}{3} e^{-t} \cos(3t - \arctan(\frac{1}{3}))$$

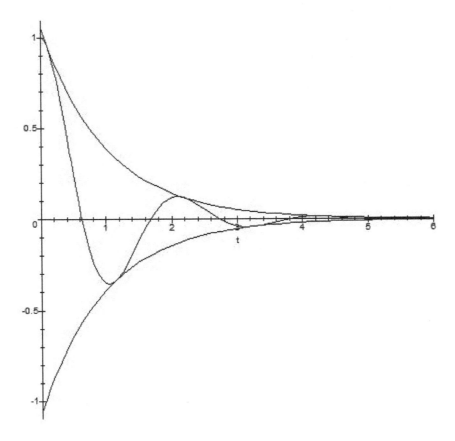

Figure 3.5: Plots of the solution of Example (3.11)

Thus, the bounding curves are

$$y = \pm \frac{\sqrt{10}}{3} e^{-t}.$$

All three plots are given in Figure (3.5).   □

Example 3.12 *A spring with spring constant 20N/m is attached to a 1kg mass with friction constant 8Ns/m. If the mass is initially position $\frac{1}{2}$ meter to the right of equilibrium and has an initial velocity of 1 m/s toward the right, determine:*

(a) when the mass will first return to the equilibrium position,

(b) the maximum displacement of the mass for $t > 0$.
(c) Use a sketch of the solution to verify your findings.

Solution: From above, we see that $m = 1$, $b = 8$ and $k = 18$. Also, we see that $b^2 - 4mk = 8^2 - 4 \cdot 1 \cdot 18 = 64 - 72 = -8$, so the spring mass system is underdamped.

The roots of the characteristic polynomial (which is $r^2 + 8r + 18 = (r+4)^2 + 2$) are $r_{1,2} = -4 \pm \sqrt{2}i$. So the general solution is

$$y(t) = c_1 e^{-4t} \cos(\sqrt{2}t) + c_2 e^{-4t} \sin(\sqrt{2}t).$$

The initial condition $y(0) = \frac{1}{2}$ yields $c_1 = \frac{1}{2}$.

Since (by the product rule)

$$y'(t) = -4c_1 e^{-4t} \cos(\sqrt{2}t) - \sqrt{2}c_1 e^{-4t} \sin(\sqrt{2}t) - 4c_2 e^{-4t} \sin(\sqrt{2}t) + \sqrt{2}c_2 e^{-4t} \cos(\sqrt{2}t),$$

we obtain (since $y'(0) = 1$)

$$1 = -4c_1 + \sqrt{2}c_2$$

Thus,

$$c_2 = \frac{3}{\sqrt{2}}$$

So

$$y(t) = \frac{1}{2} e^{-4t} \cos(\sqrt{2}t) + \frac{3}{\sqrt{2}} e^{-4t} \sin(\sqrt{2}t)$$

Solving $y(t) = 0$ yields

$$0 = e^{-4t} \left(\frac{1}{2} \cos(\sqrt{2}t) + \frac{3}{\sqrt{2}} \sin(\sqrt{2}t) \right)$$

so

$$\frac{1}{2} \cos(\sqrt{2}t) = -\frac{3}{\sqrt{2}} \sin(\sqrt{2}t)$$

$$-\frac{\sqrt{2}}{6} = \tan(\sqrt{2}t).$$

The first positive t value when this will occur will be (tangent is π-periodic)

$$t = \frac{1}{\sqrt{2}} (\pi + \arctan(-\frac{\sqrt{2}}{6})) \approx 2.057762255,$$

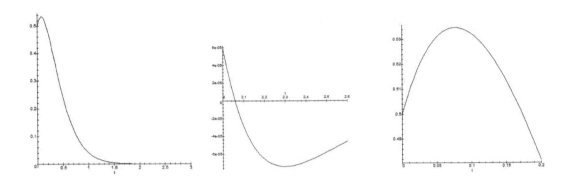

Figure 3.6: Plots of the solution of Example (3.12)

this is the answer to (a).

To solve when the maximum displacement occurs we solve $y'(t) = 0$. This is somewhat easier to do if the solution is written in phase/angle notation, which is $y(t) = e^{-4t} A \cos(\sqrt{2}t - \phi)$ where $A = \sqrt{\frac{1}{4} + \frac{9}{2}} = \frac{\sqrt{19}}{2}$ and $\phi = \arctan(\frac{\frac{3}{\sqrt{2}}}{\frac{1}{2}}) = \arctan(\frac{6}{\sqrt{2}})$ Clearly,

$$y'(t) = -4e^{-4t} A \cos(\sqrt{2}t - \phi) - \sqrt{2}e^{-4t} A \sin(\sqrt{2}t - \phi)$$

$$= e^{-4t}\left(-4\cos(\sqrt{2}t - \phi) - \sqrt{2}\sin(\sqrt{2}t - \phi)\right)$$

which is zero when

$$-4\cos(\sqrt{2}t - \phi) = \sqrt{2}\sin(\sqrt{2}t - \phi)$$

or

$$\frac{-4}{\sqrt{2}} = \tan(\sqrt{2}t - \phi)$$

Solving for t,

$$t = \frac{1}{\sqrt{2}}(\arctan(\frac{-4}{\sqrt{2}}) + \phi)$$

or

$$t = \frac{1}{\sqrt{2}}(\arctan(\frac{-4}{\sqrt{2}}) + \arctan(\frac{6}{\sqrt{2}})) \approx 0.07662176975$$

Several plots are given in Figure (3.6). Note that in order to see the first time that the mass passes through equilibrium, one needs to zoom in. The third plot zooms in on the first maxima. Notice how the first plot seems to suggest that the spring mass system is overdamped (which it is not). □

Example 3.13 *(English Units) A 16 pound weight is attached to a spring with friction constant 8lb · s/ft and spring constant 7lb/ft. Write the associated spring/mass ODE.*

Solution: In the English system, pounds are a unit of force and not mass. To convert, we use the formula from physics $F = ma$ (or sometimes given as $W = mg$) where $a = 32ft/(sec)^2$ (the acceleration due to gravity) so $16 = m32$ and $m = \frac{1}{2}$. So we obtain

$$\frac{1}{2}y'' + 8y' + 7y = 0$$

where y is measured in feet. □

Energy in an Unforced Spring Mass System

In a spring mass system, the total energy at time t is given by

$$E(t) = \frac{1}{2}m\left[y'(t)\right]^2 + \frac{1}{2}k\left[y(t)\right]^2. \tag{3.11}$$

(If mass is in kilograms, time in seconds, displacement in meters, and k in Newtons per meter, then the units are Joules (or Newton meters).

Note that for an unforced spring mass system:

$$\frac{dE}{dt} = my'(t)y''(t) + ky(t)y'(t) = y'(t)[my'' + ky]$$

so

$$\frac{dE}{dt} = my'(t)y''(t) + ky(t)y'(t) = y'(t)[-by'(t)] = -b[y'(t)]^2.$$

Note that $\frac{dE}{dt} \leq 0$ which implies that the total energy is decreasing when $y'(t) \neq 0$. The loss of energy over $[0, T]$ can be computed by

$$E(T) - E(0) = -b\int_0^T [y'(t)]^2 \, dt$$

Example 3.14 *Compute the energy loss over $[0, T]$ for a spring mass system whose mass is 1 and whose spring constant is 5 and whose motion is given by*

$$y(t) = -\frac{1}{2}e^{-2t}.$$

Solution: Clearly

$$y'(t) = e^{-2t}.$$

We know that the characteristic equation

$$r^2 + br + 5 = 0$$

plugging in $r = -2$ we get an equation for b.

$$4 - 2b + 5 = 0$$

so $b = \frac{9}{2}$.

Then

$$E(T) - E(0) = -\frac{9}{2} \int_0^T e^{-4t} \, dt$$

$$= \frac{9}{8} \, e^{-4t} \Big|_0^T = \frac{9}{8} \left(e^{-4T} - 1 \right)$$

Alternatively, we could have directly computed both $E(T)$ and $E(0)$ from formula 3.11.

$\square$

Exercises

1. A spring with spring constant $5N/m$ is attached to a 1 kg mass with friction constant $6Ns/m$. If the mass is initially position at equilibrium and has an initial velocity of 3 m/s toward the right, determine the time that the spring will be the furthest from equilibrium. What is the maximum displacement?

2. A spring with spring constant $4N/m$ is attached to a 1kg mass with friction constant $4Ns/m$. If the mass is displaced 1m to the left and has an initial velocity of 1 m/s to the right, determine if and when the mass will pass through equilibrium (t can be negative).

3. The mass of an overdamped spring system is released with a positive displacement and an initial velocity in the direction away from the equilibrium. Explain why the mass will not pass through the equilibrium position.

4. A spring mass system is underdamped with $m = 1$, $b = 2$ and $k = 10$. Initially the mass is 1 meter to the right and has an initial velocity of 2m/s away from the equilibrium.

   (a) Find the solution and write it in phase/angle notation. Use it to find the first time that the mass will pass through the equilibrium position.

   (b) Determine the velocity $y'(t)$ and express it in phase/angle notation. Use it to determine when the first local maxima will occur.

5. An overdamped spring/mass system has $m = 1$, $b = 4$ and $k = 1$ with the mass displaced 1 m to the left. Prove that if $y'(0) > 2 + \sqrt{3}$ then the mass will pass through the equilibrium position for some $t > 0$, but if $y'(0) < 2 + \sqrt{3}$ the mass will not pass through equilibrium for $t > 0$.

6. Sketch the graph of solutions with $y(0) = 1$ and $y'(0) = -1$ to the following spring/mass systems.

   (a) $m = 1$, $b = 3$ and $k = 1$

   (b) $m = 1$, $b = 2$ and $k = 1$

   (c) $m = 1$, $b = 1$ and $k = 1$

7. (a) Write and solve an ODE/IVP that models the following spring system. At equilibrium, a spring with spring constant 27 N/m suspends a mass of 3 kg. Assume that there is no friction. At time $t = 0$, the weight is displaced 1 meter from equilibrium and released (at rest).

   (b) Sketch the graph of the solution in (a).

8. Find how much energy is lost from $t = 0$ to $t = \pi$ for the solution to the spring mass system with $m = 1$ given by $y(t) = e^{-2t} \sin(3t)$.

9. A spring/mass system has $m = 1$, $b = 2$ and $k = 3$. Initially, it is displaced one unit to the left with zero initial velocity. Find a formula that solves for all times when the mass passes through equilibrium. When does it pass through the equilibrium a second time (after $t = 0$)?

10. (a) Consider $y(t) = e^{-t} \cos(2t) + 3e^{-t} \sin(2t)$. Write a spring mass ODE that this solves.

   (b) Plot $y(t) = e^{-t} \cos(2t) + 3e^{-t} \sin(2t)$ (use technology and printout the graph)

   (c) Put the above in $y(t) = Ae^{-t} \cos(2t - \phi)$ phasor notation

   (d) On the plot in (b) also plot $y = Ae^{-t}$ and $y = -Ae^{-t}$

   (e) Find all times that this curve hits $y = Ae^{-t}$

   (f) Find all times when $y' = 0$ for $y(t) = e^{-t} \cos(2t) + 3e^{-t} \sin(2t)$. Are these the same as (e)? Why not?

3.5 Undetermined Coefficients

In the previous section we learned how to solve all possible homogeneous second order linear ODE with constant coefficients, written as:

$$ay'' + by' + cy = 0$$

for constants a, b, c.

In this section we will solve

$$ay'' + by' + cy = f(t)$$

for a specific class of functions $f(t)$.

Finding one solution to the nonhomogeneous

We motivate this method with an example:

$$y'' + 5y' + 6y = 4e^t$$

Our goal is to find one solution to this differential equation. In particular, we ask ourselves, what kind of functions might possibly solve this nonhomogeneous differential equation? Clearly, polynomials or trigonometric functions will not do. The only possible solution would be a function that involves e^t so we make a reasonable guess at the form of a solution by considering functions of the form $y = Ae^t$. Our strategy will be to plug this function into the DE and solve for the constant A to force the function to solve the differential equation.

After plugging this function into the left side of DE we obtain

$$y'' + 5y' + 6y = Ae^t + 5Ae^t + 6Ae^t = 12Ae^t$$

So in order to solve the DE (to match the right hand side) we must choose $12A = 4$, or $A = \frac{1}{3}$ in order to obtain $4e^t$. Thus, we have found that $y = \frac{1}{3}e^t$ solves the DE.

We seem to have been extremely lucky in the previous example. Since the function $y = Ae^t$ has derivatives that have the same form as the function itself. If the function $f(t)$ had been $4\cos(2t)$ then we would have made a guess of the form $y = A\cos(2t) + B\sin(2t)$. The sine term needs to be included in our guess since the derivative of the cosine term will involve sines. We solve the problem below:

Example 3.15 *Find one solution to the DE*

$$y'' + 5y' + 6y = 4\cos(2t)$$

Solution: We guess at the form of a solution $y = A\cos(2t) + B\sin(2t)$.

$$y' = -2A\sin(2t) + 2B\cos(2t)$$

$$y'' = -4A\cos(2t) - 4B\sin(2t).$$

Plugging into the left-hand DE we obtain:

$$-4A\cos(2t) - 4B\sin(2t) + 5\left(-2A\sin(2t) + 2B\cos(2t)\right) + 6\left(A\cos(2t) + B\sin(2t)\right)$$

$$= (2A + 10B)\cos(2t) + (-10A + 2B)\sin(2t)$$

matching coefficients, we want the right side to be $4\cos(2t) + 0\sin(2t)$, we obtain

$$2A + 10B = 4 \quad \text{and} \quad -10A + 2B = 0.$$

By the second equation, we see $B = 5A$. Plugging this back into the first equation gives

$2A + 50B = 4$ or $A = \frac{1}{13}$. Therefore, $B = \frac{5}{13}$.

So, we find that

$$y = \frac{1}{13}\cos(2t) + \frac{5}{13}\sin(2t)$$

solves the DE. $\qquad\qquad\qquad\qquad\qquad\qquad\qquad\qquad\qquad\qquad\qquad\quad\square$

As you may have guessed, this method works if $f(t)$ is a function whose derivatives have forms that do not get infinitely complicated. For instance if $f(t) = \frac{1}{t}$ then this method will not work, since to account for all possible derivatives, our guess would need to be of the form:

$$y = A_1\left(\frac{1}{t}\right) + A_2\left(\frac{1}{t^2}\right) + A_3\left(\frac{1}{t^3}\right) + \dots$$

Example 3.16 *Find one solution to the DE*

$$y'' + 5y' + 6y = 12t^2$$

Solution: We guess at the form of a solution $y = At^2 + Bt + C$. We need the linear and constant terms since if we only have a quadratic term, both $5y'$ will involve a linear term, which is not on the right hand side, so we need to include linear and constant term in our guess in order to try to cancel this linear term.

We compute

$$y' = 2At + B$$

$$y'' = 2A.$$

Plugging into the left-hand DE we obtain:

$$2A + 5\left(2At + B\right) + 6\left(At^2 + Bt + C\right)$$

$$= (6A)t^2 + (10A + 6B)t + 2A + 5B + 6C$$

matching coefficients, we want the right side to be $1t^2 + 0t + 0$, we obtain

$$6A = 12 \quad \text{and} \quad 10A + 6B = 0 \quad \text{and} \quad 2A + 5B + 6C = 0$$

solving, we obtain

$$A = 2 \quad \text{and} \quad B = -\frac{10}{3} \quad \text{and} \quad C = \frac{19}{9}$$

So

$$y = 2t^2 - \frac{10}{3}t + \frac{19}{9}$$

solves the DE. $\qquad \square$

We provide one more example to illustrate the need make a guess whose derivatives are also of the form of the original guess.

Example 3.17 *Find one solution to the DE*

$$y'' + 4y' + y = 2xe^{-x}$$

Solution: We guess at the form of a solution $y = Axe^{-x} + Be^{-x}$. Again, we need our guess to have the property that the derivatives of our guess have the same form as our guess itself.

$$y' = Ae^{-x} - Axe^{-x} - Be^{-x}$$

$$y'' = -Ae^{-x} - Ae^{-x} + Axe^{-x} + Be^{-x}.$$

Plugging into the DE, we obtain:

$$-2Ae^{-x} + Be^{-x} + Axe^{-x} + 4\left(Ae^{-x} - Axe^{-x} - Be^{-x}\right) + Axe^{-x} + Be^{-x}$$

$$= -2Axe^{-x} + (2A - 2B)e^{-x},$$

which we wish to set equal to

$$2xe^{-x} + 0e^{-x}$$

Matching coefficients,

$$-2A = 2 \quad \text{and} \quad 2A - 2B = 0$$

so

$$A = -1 \quad \text{and} \quad B = -1$$

and $y = -xe^{-x} - e^{-x}$. $\square$

This method could conceivably break down if the general solution to the homogeneous DE has the same form (or parts of the same form) as the guess function as demonstrated by the following example:

Example 3.18 *Show that $y = Ae^{-5x}$ cannot be a solution to the equation*

$$y'' + 6y' + 5y = 4e^{-5x}.$$

Solution: Consider $y = Ae^{-5x}$. We know $y' = -5Ae^{-5x}$ and $y'' = 25Ae^{-5x}$.
   Plugging into the left-hand side of the DE we obtain

$$25Ae^{-5x} + 6\left(-5Ae^{-5x}\right) + 5\left(Ae^{-5x}\right) = 0$$

The trouble here is that our guess actually coincides with a term from the homogeneous general solution so there is no possible choice of A to satisfy the DE.

Recall that the characteristic polynomial of the DE is

$$r^2 + 6r + 5 = (r+1)(r+5)$$

which has roots $r_1 = -1$ and $r_2 = -5$.

So the general solution to the homogeneous is

$$y_{homog} = c_1 e^{-x} + c_2 e^{-5x}$$

whose second term coincides with the that guess we made. □

In this case, we can recover the solution to the DE by multiplying the original guess by x. (This may seem out of the blue, but when one considers the effects of the product rule, this scheme makes better sense).

Example 3.19 *Show that there is a solution of the form $y = Axe^{-5x}$ to the differential equation*

$$y'' + 6y' + 5y = 4e^{-5x}.$$

Solution: We know $y' = -5Axe^{-5x} + Ae^{-5x}$ and $y'' = 25Axe^{-5x} - 5Ae^{-5x} - 5Ae^{-5x}$. Plugging into the left-hand side of the DE we obtain

$$25Axe^{-5x} - 10Ae^{-5x} + 6\left(-5Axe^{-5x} + Ae^{-5x}\right) + 5\left(Axe^{-5x}\right)$$

$$= -4Ae^{-5x},$$

so if $A = -1$ then the left-side will equal the right-hand side. So $y = -xe^{-5x}$ solves the DE. □

In the previous example, note that we do not need a term of the form Be^{-5x} since plugging into the DE will just result in zero as we saw before. In general practice, if a term of your guess coincides with a term from the homogeneous DE, one multiplies by the smallest power of the independent variable and adjusts the guess. In the next example, we are forced to multiply by the square of the variable.

Example 3.20 *Find one to the differential equation*

$$y'' - 8y' + 16y = e^{4t}.$$

Solution: Notice that the guess

$$y = Ae^{4t}$$

unfortunately coincides with one term in the homogeneous solution

$$y = c_1 e^{4t} + c_2 t e^{4t}.$$

So we have no hope of finding an A to solve the nonhomogeneous. To remedy, we multiply our guess by t^2 (multiplying by t will still not work).

So we will work with

$$y = At^2 e^{4t}$$

We know $y' = 2At e^{4t} + 4At^2 e^{4t}$ and $y'' = 2Ae^{4t} + 8At e^{4t} + 8At e^{4t} + 16At^2 e^{4t}$. Plugging into the left-hand side of the DE we obtain

$$2Ae^{4t} + 16At e^{4t} + 16At^2 e^{4t} - 8\left(2At e^{4t} + 4At^2 e^{4t}\right) + 16\left(At^2 e^{4t}\right)$$

$$= 2Ae^{4t},$$

so if $A = \frac{1}{2}$ then the left-side will equal the right-hand side. So $y = \frac{1}{2}t^2 e^{4t}$ solves the DE. $\square$

Below is a chart to assist in determining the appropriate guess:

Undetermined Coefficients Method

The nonhomogeneous second order linear DE with constant coefficients

$$ay'' + by' + cy = f(t) \tag{3.12}$$

has a solution of the form:

f(t)	Guess
$constant$	A
e^{rt}	Ae^{rt}
$\cos(kt)$	$A\cos(kt) + B\sin(kt)$
$\sin(kt)$	$A\cos(kt) + B\sin(kt)$
$a_n t^n + a_{n-1}t^{n-1} + ... + a_1 t + a_0$	$A_n t^n + A_{n-1}t^{n-1} + ... + A_1 t + A_0$
$(n^{th} \ degree \ poly) \cdot \cos(kt)$ $+(n^{th} \ degree \ poly) \cdot \sin(kt)$	$(A_n t^n + ... + A_1 t + A_0)\cos(kt)$ $+(B_n t^n + ... + B_1 t + B_0)\sin(kt)$
$(n^{th} \ degree \ poly) \cdot e^{rt}$	$(A_n t^n + ... + A_1 t + A_0)e^{rt}$
$(n^{th} \ degree \ poly) \cdot e^{rt} \cdot \cos(kt)$ $+(n^{th} \ degree \ poly) \cdot e^{rt} \cdot \sin(kt)$	$(A_n t^n + ... + A_1 t + A_0)e^{rt}\cos(kt)$ $+(B_n t^n + ... + B_1 t + B_0)e^{rt}\sin(kt)$

The unknown constants in the guess are obtained by plugging the guess into the DE and solving for the coefficients.

The exception is when terms from the above guess coincide with the homogeneous solution to the DE. In these cases, the guess term needs to be multiplied by the smallest power of t so that the guess no longer has terms that coincide with the general solution of the homogeneous.

Finding general solutions to the nonhomogeneous

Next, we describe how we can use one solution to the nonhomogeneous differential equation to generate the general solution to the nonhomogeneous DE. The following theorem concerning linear DEs allows us to do this. This theorem is sometimes called the superposition principle, since one solution can be super-imposed onto the other. Note that this theorem holds for general linear second order ODE (which includes the case of constant coefficients).

Superposition Principle

Suppose $y_1(t)$ solves the linear DE

$$a(t)y'' + b(t)y' + c(t)y = f(t) \tag{3.13}$$

and that $y_2(t)$ solves the linear DE

$$a(t)y'' + b(t)y' + c(t)y = g(t) \tag{3.14}$$

Then $Y(t) = y_1(t) + y_2(t)$ solves

$$a(t)y'' + b(t)y' + c(t)y = f(t) + g(t) \tag{3.15}$$

Proof: Consider

$$Y(t) = y_1(t) + y_2(t).$$

Differentiating and using the fact that the derivative of the sum equals the sum of the derivatives:

$$Y'(t) = y_1'(t) + y_2'(t).$$

$$Y''(t) = y_1''(t) + y_2''(t).$$

So plugging into the left side of DE (3.18), we obtain

$$a(t)Y'' + b(t)Y' + c(t)Y$$

$$= a(t)(y_1'' + y_2'') + b(t)(y_1' + y_2') + c(t)(y_1 + y_2)$$

$$= a(t)y_1'' + b(t)y_1' + c(t)y_1 + a(t)y_2'' + b(t)y_2' + c(t)y_2$$

$$= f(t) + g(t).$$

So $Y(t)$ solves DE (3.18). $\square$

Note that in the above theorem, we assume that $Y(t) = y_1(t) + y_2(t)$ exists, meaning that we assume that y_1 and y_2 have compatible domains.

The following result is a straightforward consequence of the superposition principle:

Using Undetermined Coefficients for General Solutions

Suppose $y_{homog}(t)$ is the general solution to the homogeneous linear DE

$$a(t)y'' + b(t)y' + c(t)y = 0 \qquad (3.16)$$

and that $y_P(t)$ is any one solution to the nonhomogeneous linear DE

$$a(t)y'' + b(t)y' + c(t)y = f(t) \qquad (3.17)$$

Then $Y(t) = y_{homog}(t) + y_P(t)$ is the general solution for

$$a(t)y'' + b(t)y' + c(t)y = f(t) \qquad (3.18)$$

This allows us to find general solutions to nonhomogeneous DEs. First, we find the general solution to the associated homogeneous DE and add it to one solution to the nonhomogeneous DE.

Example 3.21 *Find the general solution to the differential equation*

$$y'' - 2y' + y = 3e^{4t}.$$

Solution: As before, we can find one solution to the nonhomogeneous by making a guess

$$y = Ae^{4t}$$

Plugging into the left-side of the DE, we obtain

$$16Ae^{4t} - 8Ae^{4t} + Ae^{4t} = 9Ae^{4t}$$

So to match the right side of the DE $9Ae^{4t} = 3e^{4t}$, so we take $A = \frac{1}{3}$ or we see that one solution to the nonhomogeneous DE is

$$y_P(t) = \frac{1}{3}e^{4t}$$

Next, we realize that the general solution to

$$y'' - 2y' + y = 0$$

is given by

$$y_{homog}(t) = c_1 e^t + c_2 t e^t$$

Thus, by the superposition principle,

$$y(t) = y_{homog}(t) + y_P(t) = c_1 e^t + c_2 t e^t + \frac{1}{3} e^{4t}$$

is the general solution to the nonhomogeneous DE.   □

Example 3.22 *Find the solution to the initial value problem*

$$y'' - 3y' + 2y = \sin(2t), \quad y(0) = 0, \quad y'(0) = 0$$

Solution: As before, we can find one solution to the nonhomogeneous by making a guess

$$y = A \sin(2t) + B \cos(2t).$$

We obtain

$$y' = 2A \cos(2t) - 2B \sin(2t)$$

$$y'' = -4A \sin(2t) - 4B \cos(2t)$$

and plug into the lefthand side of the DE:

$$(-4A - 6B + 2A) \sin(2t) + (-4B + 6A + 2B) \cos(2t)$$
$$= (-2A - 6B) \sin(2t) + (6A - 2B) \cos(2t)$$

Matching coefficients, we wish to obtain

$$1 \sin(2t) + 0 \cos(2t)$$

to obtain:

$$-2A - 6B = 1 \quad \text{and} \quad 6A - 2B = 0.$$

Multiplying the first equation by 3, we obtain

$$-6A - 18B = 3 \quad \text{and} \quad 6A - 2B = 0.$$

and adding

$$-20B = 3$$

or

$$B = -\frac{3}{20}$$

and

$$6A = 2B$$

implies that

$$A = -\frac{1}{20}$$

Thus,

$$y_P(t) = -\frac{1}{20}\sin(2t) - \frac{3}{20}\cos(2t).$$

Is one solution to the nonhomogeneous solution.

The homogeneous solution is

$$y_{homog}(t) = c_1 e^t + c_2 e^{2t}$$

Hence the general solution to the nonhomogeneous DE is

$$y(t) = c_1 e^t + c_2 e^{2t} - \frac{1}{20}\sin(2t) - \frac{3}{20}\cos(2t)$$

Solving for the constants:

$$y(0) = 0$$

implies that

$$0 = c_1 + c_2 - \frac{3}{20}$$

also,

$$y'(t) = c_1 e^t + 2c_2 e^{2t} - \frac{1}{10}\cos(2t) + \frac{3}{10}\sin(2t)$$

so

$$0 = y'(0) = c_1 + 2c_2 - \frac{1}{10}$$

So

$$c_1 + c_2 = \frac{3}{20} \quad \text{and} \quad c_1 + 2c_2 = \frac{1}{10}$$

subtracting, we obtain

$$c_2 = -\frac{1}{10}$$

and

$$c_1 = \frac{5}{20}$$

So,

$$y(t) = \frac{5}{20}e^t - \frac{1}{10}e^{2t} - \frac{1}{20}\sin(2t) - \frac{3}{20}\cos(2t)$$

is the particular solution to the nonhomogeneous initial value problem.   □

Note that the superposition principle allows us to split up a more complicated problem into smaller pieces. For instance, if we were trying to find one solution to

$$y'' - 3y' + 2y = \sin(2t) + e^{-t} + t^2 + 8t - 6$$

we could find a solution y_1 to

$$y'' - 3y' + 2y = \sin(2t),$$

a solution y_2 to

$$y'' - 3y' + 2y = e^{-t},$$

a solution y_3 to

$$y'' - 3y' + 2y = t^2 + 8t - 6,$$

and $y_1 + y_2 + y_3$ would solve

$$y'' - 3y' + 2y = \sin(2t) + e^{-t} + t^2 + 8t - 6.$$

Exercises

Find ONE solution for each of the DEs

1. $y'' + 4y' + 4y = 2e^t$

2. $y'' + 3y' + 2y = t^2 - 4t$

3. $y'' + 6y' - 7y = \cos(4t)$

4. $z'' + 4z' + z = 4$

5. $z'' - z = 6e^x$

6. $z'' + 2z' = 6\sin t$

7. $z'' + 6z' + 5z = 7e^{-t}$

8. $z'' + 6z' + 5z = te^{-t}$

9. $z'' + 6z' + 5z = \sin(t + \frac{\pi}{4})$ (Use a trig formula)

10. $z'' + 6z' + 5z = \cosh(2t)$

Find general solutions for each of the DEs

11. $y'' + 4y' + 4y = \sin x$

12. $y'' + 3y' + 2y = e^t$

13. $y'' + 6y' - 5y = t^2 + 1$

14. $z'' + 4z' + z = t\sin t$

15. $z'' - z = t^3$

16. $z'' + z' = 6$

17. $z'' + 7z' + 6z = e^{-t}$

18. $z'' + 7z' + 3z = \cos(t + \frac{\pi}{3})$ (Use a trig formula)

19. $z'' + 7z' + 6z = e^{2t+4}$ (Expand the exponential)

Find the particular solution to the initial value problems

20. $y'' + 4y' + 4y = 2e^t, \quad y(0) = 1, \quad y'(0) = 0$

21. $y'' + 3y' + 2y = t^2 - 4t, \quad y(0) = 0, \quad y'(0) = 0$

22. $y'' + 6y' - 7y = \cos(4t), \quad y(0) = 0, \quad y'(0) = 0$

23. $z'' + 4z' + z = 4, \quad z(0) = 0, \quad z'(0) = -1$

3.6 Application-Forced Spring Mass Systems and Resonance

In this section we introduce an external force that acts on the mass of the spring in addition to the other forces that we have been considering. For example, suppose that the mass of a spring/mass system is being pushed (or pulled) by an additional force (perhaps the spring is mechanically driven or is being acted upon by magnetic forces). We will call this net external force $F_{external}$ and allow it to vary over time, that is $F_{external} = F(t)$.

As before,

$$F_{total} = -by' - ky + F_{external}$$

and by Newton's Law

$$my'' = -by' - ky + F_{external}.$$

So we obtain the nonhomogeneous ODE:

$$my'' + by' + ky = F(t).$$

As we saw earlier, this nonhomogeneous can be solved by using the principle of superposition, where $y_{homog}(t)$ is the general solution to the associated homogeneous and $y_P(t)$ is any one solution to the nonhomogeneous DE.

From the previous section, so long as $b > 0$, then $\lim_{t \to \infty} y_{homog}(t) = 0$, so the long-term behavior of any solution to the nonhomogeneous will be determined by the behavior of $y_P(t)$. In such problems, the associated homogeneous solution $y_{homog}(t) \to 0$ is called transient part of the solution (since it dies away) and $y_P(t)$ is called the *steady state solution* since it determines the long-term behavior.

Example 3.23 *A spring with spring constant 4N/m is attached to a 1kg mass with friction constant 4Ns/m is forced to the right by a constant force of 2N. Find the steady state solution.*

Solution: In light of the discussion above, we need only find $y_P(t)$ which we can obtain by undetermined coefficients on the non homogeneous ODE

$$y'' + 4y' + 4y = 2$$

to obtain $y_P(t) = \frac{1}{2}$ meter. So no matter what the initial conditions are, the mass will limit to a displacement $\frac{1}{2}$ meter to the right. $\square$

Example 3.24 *A spring with spring constant 4N/m is attached to a 1kg mass with friction constant 4Ns/m is forced periodically by a constant force of 2 cos(t)N. (a) Find the steady state solution and express it in phase/angle notation.*

(b) Find the particular solution that satisfies $y(0) = 1$ and $y'(0) = 2$, and verify that the graph limits to the steady state solution.

Solution: In light of the discussion above, we need only find $y_P(t)$ which we can obtain by undetermined coefficients on the non homogeneous ODE

$$y'' + 4y' + 4y = 2\cos t.$$

We use undetermined coefficients on the form: $y_P(t) = A\cos t + B\sin t$ and obtain

$$y'' + 4y' + 4y = -A\cos t - B\sin t - 4A\sin t + 4B\cos t + 4A\cos t + 4B\sin t$$

$$= (3A + 4B)\cos t + (3B - 4A)\sin t$$

which we set equal to $2\cos t + 0\sin t$ so

$$3A + 4B = 2 \quad and \quad 3B - 4A = 0,$$

so

$$12A + 16B = 8 \quad and \quad 9B - 12A = 0,$$

which, when added, yields
$$25B = 8$$

so $B = \frac{8}{25}$ and $A = \frac{6}{25}$.

So
$$y_P(t) = \frac{1}{25}(6\cos t + 8\sin t)$$

which can be expressed as

$$y_P(t) = \frac{\sqrt{36 + 64}}{25}(\cos(t - \arctan(\frac{4}{3}))) = \frac{2}{5}(\cos(t - \arctan(\frac{4}{3}))).$$

To solve part (b), notice that the general solution to the DE is given by

$$y(t) = y_{homog}(t) + y_P(t) = c_1 e^{-2t} + c_2 t e^{-2t} + \frac{1}{25}(6\cos t + 8\sin t)$$

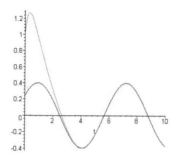

Figure 3.7: Plots of the solution and the steady state solution from Example (3.24)

so the particular solution that we seek satisfies $y(0) = c_1 + \frac{6}{25} = 1$ or $c_1 = \frac{19}{25}$ and

$$y'(t) = -2(\frac{19}{25})e^{-2t} - 2c_2te^{-2t} + c_2e^{-2t} + \frac{1}{25}(8\cos t - 6\sin t)$$

so since $y'(0) = 2$ we obtain $2 = -\frac{38}{25} + c_2 + \frac{8}{25}$ $c_2 = \frac{80}{25} = \frac{16}{5}$. So the particular solution we seek is

$$y(t) = y_{homog}(t) + y_P(t) = \frac{19}{25}e^{-2t} + \frac{16}{5}te^{-2t} + \frac{1}{25}(6\cos t + 8\sin t).$$

The plots of the steady state solution and the particular solution are given in figure (3.6), notice how the solution limits to the steady state. $\square$

Resonance

In this section we look at a particular phenomenon called resonance. In principle, anyone who has ever pushed a child on a swing is familiar with this concept. A child swinging on a swing will oscillate back and forth with a given frequency. In order to push the child effectively on the swing, the frequency of the pushes needs to coincide with the frequency of the swing otherwise, you will be pushing when the child is swinging toward you. This principle is also what is responsible for the ability of an opera singer to shatter a champagne glass, the oscillatory forces that are capable of destroying a bridge, or exciting molecules at their natural frequency in microwave ovens. More formally, a spring/mass system exhibits *resonance* if the steady state solution obtained periodically forcing the system in an oscillatory manner results in a greater maximum displacement than the steady state solution obtained by forcing the system with a constant force F_0, where F_0 is the amplitude of the oscillatory forcing.

Example 3.25 *Compare system (a)* $y''+y'+6y = 2$ *with system (b)* $y''+y'+6y = 2\sin 3t$. *Which has a larger amplitude for the steady state solution?*

Solution: (a) The steady state solution to

$$y'' + y' + 6y = 2$$

is $y(t) = \frac{1}{3}$. (Use undetermined coefficients on $y = A$, and solve for A.)
  (b) The steady state solution to

$$y'' + y' + 6y = 2\sin 3t$$

is obtained by plugging $y = A\cos 3t + B\sin 3t$ into the left side of the DE, we obtain

$$-9A\cos 3t - 9B\sin 3t - 3A\sin 3t + 3B\cos 3t + 6A\cos 3t + 6B\sin 3t$$

$$= (-3A + 3B)\cos 6t + (3A - 3B)\sin 6t$$

so

$$-3A + 3B = 0 \quad \text{and} \quad -3A - 3B = 2$$

solving, we get

$$-6A = 2$$

so

So $A = -\frac{1}{3}$ and $B = -\frac{1}{3}$ which implies that the steady state solution is $y(t) = -\frac{1}{3}\sin 3t - \frac{1}{3}\cos 3t$ which can be rewritten in phase/angle form as $y(t) = \frac{\sqrt{2}}{3}\cos(3t - (\frac{\pi}{4} + \pi))$
  Clearly, the amplitude obtained in (b) is larger than the one obtained in (a) as Figure (3.11) demonstrates, so the system exhibits resonance.   □

Sinusoidal Forcing

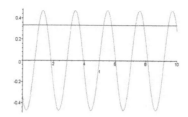

Figure 3.8: Plots of the two steady state solutions from Example (3.25)

Suppose that a spring/mass system with spring constant $k > 0$ is attached to a mass of $m > 0$ kilograms with with friction constant $b > 0$. We wish to examine when a sinusoidal forcing function of the form $F_0 \cos(\omega t - \phi)$ produces a steady state solution with a larger amplitude than the steady state solution obtained by forcing with constant force of $F_0 > 0$.

As before, by the method of undetermined coefficients,

$$my'' + by' + ky = F_0$$

has a steady state solution of $y(t) = F_0/k$.

Next, we use undetermined coefficients to solve

$$my'' + by' + ky = F_0 \cos(\omega t - \phi)$$

with

$$y = A\cos(\omega t - \phi) + B\sin(\omega t - \phi)$$

to obtain

$$my'' = -\omega^2 Am\cos(\omega t - \phi) - Bm\omega^2 \sin(\omega t - \phi)$$
$$by' = \omega Bb\cos(\omega t - \phi) - Ab\omega \sin(\omega t - \phi)$$
$$ky = Ak\cos(\omega t - \phi) + Bk\sin(\omega t - \phi)$$

So matching coefficients:

$$-\omega^2 Am + Bb\omega + Ak = F_0 \quad \text{and} \quad - Bm\omega^2 - Ab\omega + Bk = 0.$$

The second equation is easily solved as

$$B = \frac{Ab\omega}{k - m\omega^2}.$$

Therefore, by the first equation:

$$-\omega^2 Am + b\omega \left(\frac{Ab\omega}{k - m\omega^2} \right) + Ak = F_0.$$

Solving this for A yields

$$A = \frac{F_0(k - m\omega^2)}{b^2\omega^2 + (k - m\omega^2)^2}$$

and

$$B = \frac{F_0 b\omega}{b^2\omega^2 + (k - m\omega^2)^2}.$$

So the steady state solution will have amplitude

$$\sqrt{A^2 + B^2} = \sqrt{\frac{F_0^2}{b^2\omega^2 + (k - m\omega^2)^2}}$$

This exceeds $\frac{F_0}{k}$ precisely when

$$k^2 > b^2\omega^2 + (k - m\omega^2)^2.$$

or

$$\omega^2 \left(-m^2\omega^2 + (2mk - b^2) \right) > 0$$

or

$$(2mk - b^2) > m^2\omega^2.$$

The above inequality implies that $b^2 - 2mk < 0$, since the term on the right hand side is clearly positive. Moreover, so long as

$$-\frac{\sqrt{2km - b^2}}{m} < \omega < \frac{\sqrt{2km - b^2}}{m}$$

then forcing with a periodic frequency ω produces a steady state with larger amplitude than the steady state produced by constant forcing. The interval $(0, \frac{\sqrt{2km-b^2}}{m})$ is called the **interval of resonance**. It is also true that any frequency in $(-\frac{\sqrt{2km-b^2}}{m}, 0)$ will yield a response larger in amplitude than constant forcing, but we will restrict our attention to positive frequencies.

To derive the **optimal** forcing frequency, we minimize the function $G(\omega) = b^2\omega^2 + (k - m\omega^2)^2$ with respect to ω (thereby, maximizing $= \sqrt{\frac{F_0^2}{b^2\omega^2 + (k - m\omega^2)^2}}$).

Taking a derivative, we obtain

$$G'(\omega) = 2b^2\omega + 2(k - m\omega^2)(-2m\omega)$$

and obtain

$$2b^2 - 4m(k - m\omega^2) = 0$$

or

$$\omega^2 = \frac{4mk - 2b^2}{4m^2}$$

or

$$\omega = \frac{\sqrt{4mk - 2b^2}}{2m} = \frac{\sqrt{2mk - b^2}}{\sqrt{2}m}.$$

In the exercises, you will be asked to verify that this is indeed a maximum by using the second derivative test. We summarize our work below:

Resonance in Sinusoidally Forced Spring/Mass Systems

The sinusoidally forced spring mass system

$$my'' + by' + ky = F_0\cos(\omega t - \phi) \tag{3.19}$$

exhibits resonance when $b^2 - 2mk < 0$ (we call such a system lightly damped). In particular, For any non-zero ω inside:

$$-\frac{\sqrt{2km - b^2}}{m} < \omega < \frac{\sqrt{2km - b^2}}{m},$$

the steady state solution of the periodically forced system exceeds the steady state solution of the system forced by a constant. The optimal forcing frequency (called **the resonance frequency**) is

$$\omega = \pm\frac{\sqrt{2mk - b^2}}{\sqrt{2}m} \tag{3.20}$$

Not all spring mass systems exhibit resonance. In particular, for a spring mass system that is forced by a sinusoidal function to exhibit resonance, it must be lightly damped, meaning $b^2 - 2mk < 0$. Note that lightly damped implies that

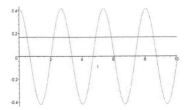

Figure 3.9: Steady state solutions of a spring mass system forced at its resonant frequency as well as being forced by a constant force of 1.

the system is underdamped since $b^2 - 4mk < b^2 - 2mk$ we have $b^2 - 2mk < 0$ implies that $b^2 - 4mk < 0$.

Example 3.26 *Compute the frequencies ω for which*

$$y'' + y' + 6y = \cos(\omega t - \phi)$$

produces resonance. Also, find the resonance frequency and plot the steady state solution when the system is forced at that frequency.

Solution: Note also that we are forcing the system sinusoidally with $F_0 = 1$. Since $m = 1$, $b = 1$ and $k = 6$, we see that $b^2 - 2mk = 1 - 12 < 0$ so the system is lightly damped and exhibits resonance.

In particular, for positive frequencies ω satisfying

$$0 < \omega < \sqrt{11},$$

the steady state solution of the periodically forced system exceeds the steady state solution of the system forced by a constant. (Note in example (3.25) since $0 < 3 < \sqrt{11}$, the steady state solution of the periodically driven system exceeds that of the system driven by a constant.)

The optimal frequency to force the system (i.e. the resonance frequency) occurs at $\omega = \frac{\sqrt{11}}{\sqrt{2}}$ and the maximum amplitude is given by

$$\text{Max Amplitude} = \sqrt{\frac{1}{11/2 + (6 - 11/2)^2}} = \frac{1}{\sqrt{\frac{23}{4}}} = \frac{2}{\sqrt{23}} \approx 0.417028828. \quad \square$$

Pure Mathematical Resonance

An interesting phenomenon occurs in a spring mass system that has no damping, i.e. $b = 0$. In particular, the steady state solution obtained by forcing at the resonance frequency is unbounded.

From the formula for the optimal resonance frequency above, we see that $\omega = \sqrt{\frac{k}{m}}$.

If we solve

$$my'' + ky = F_0 \cos(\sqrt{\frac{k}{m}}t - \phi) = F_0 \cos\phi \cos(\sqrt{\frac{k}{m}}t) - F_0 \sin\phi \sin(\sqrt{\frac{k}{m}}t)$$

then we see that the steady state solution (due to interaction with the homogenous solution) will be of the form

$$y(t) = At\cos(\sqrt{\frac{k}{m}}t) + Bt\sin(\sqrt{\frac{k}{m}}t)$$

which can, in turn, be written in phase/angle form as

$$y(t) = t\sqrt{A^2 + B^2}\cos(\sqrt{\frac{k}{m}}t - \phi).$$

This solution is clearly unbounded as $t \to \infty$ and results in (wild) oscillations with amplitude going to infinity. This occurrence is called **pure mathematical resonance,** and although it cannot occur in an actual spring/mass system, the concept is relevant to systems with extremely light damping.

Example 3.27 *Plot the steady state solution to*

$$y'' + 4y = \cos(\omega t),$$

where ω is the resonance frequency.

Solution: As above the resonance frequency by Equation (3.20) is $\omega = \sqrt{\frac{k}{m}}$. So $\omega = 2$. Applying undetermined coefficients to:

$$y'' + 4y = \cos(2t),$$

we obtain $y(t) = At\cos(2t) + Bt\sin(2t)$ yielding:

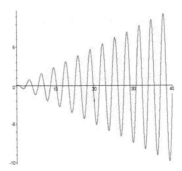

Figure 3.10: Steady state solutions associated with pure mathematical resonance.

$$y' = A\cos(2t) - 2At\sin(2t) + B\sin(2t) + 2Bt\cos(2t)$$

$$y'' = -2A\sin(2t) - 2A\sin(2t) - 4At\cos(2t) + 2B\cos(2t) + 2B\cos(2t) - 4Bt\sin(2t)$$

So $y'' + 4y = 4B\cos(2t) - 4A\sin(2t)$

and so $A = 0$ and $B = \frac{1}{4}$.

Thus the steady state solution is $y(t) = \frac{1}{4}t\cos(2t)$ which is plotted in Figure (3.10)   □

Exercises

1. Find the steady state solutions to

   (a)
   $$y'' + 2y' + 10y = 4$$

   and

   (b)
   $$y'' + 2y' + 10y = 4\cos(2t).$$

   Compare the amplitude of the steady state (b) with the steady state of (a).

   (c) Is resonance possible with this spring/mass system?

2. Find the amplitude of the steady state solution to

   $$y'' + 2y' + 3y = 4\cos(2t).$$

   Does the associated system exhibit resonance? If so, compute the resonance frequency.

3. For
   $$y'' + 2y' + 10y = 4\cos(\omega t),$$

   graph the steady state solutions for specific values of ω where ω is:

   (a) $\omega = 0$ (i.e. the constant solution).

   (b) the optimal resonance frequency.

   (c) What is the amplitude of the steady state?

4. For
   $$y'' + y' + y = 2\cos(\omega t),$$

   graph the steady state solutions for specific values of ω where ω is:

   (a) $\omega = 0$ (i.e. the constant solution).

   (b) the optimal resonance frequency.

   (c) What is the amplitude of the steady state?

5. The spring/mass system

$$y'' + 1y' + 10y = 400\cos(\omega t)$$

has the mass initially at equilibrium and at rest and is forced at the optimal resonance frequency. If the spring can tolerate a displacement of at most $y = 60$ units, when will the spring break with initial conditions $y(0) = 0$ and $y'(0) = 0$? [Hint: Plot the particular solution, not the steady state.]

6. For

$$y'' + 10y = 4\cos(\omega t),$$

plot the steady state solutions for specific values of ω where ω is the optimal resonance frequency. If the spring can tolerate a displacement of at most $y = 100$ units, when will the spring break with initial conditions $y(0) = 0$ and $y'(0) = 0$?

7. Suppose we wish to maximize the amplitude of the velocity $q'(t)$ for the steady state solution of $my'' + by' + ky = F_0\cos(\omega t - \phi)$. Show that $\omega = \sqrt{\frac{k}{m}}$ is the frequency that maximizes the amplitude of $q'(t)$ for the steady state solution. [You may use the calculations made earlier in this section.]

8. Suppose we wish to maximize the acceleration $q''(t)$ for the steady state solution to $my'' + by' + ky = F_0\cos(\omega t - \phi)$. Find when there is an optimal ω and give its formula. (Hint: There will be an inequality that must hold similar to $b^2 - 2mk < 0$).

9. Verify that $G(\omega) = b^2\omega^2 + (k - m\omega^2)^2$ has a maximum at $\omega = \frac{\sqrt{4mk - 2b^2}}{2m} = \frac{\sqrt{2mk - b^2}}{\sqrt{2}m}$ by using the second derivative test.

3.7 Application-LRC Circuits

The charge in an LRC-circuit (or more commonly called an RLC-circuit) can also be modeled by a second order linear DE with constant coefficients. The DE is

$$Lq'' + Rq' + \frac{1}{C}q = 0.$$

Here $q(t)$ is charge at some specified point in the circuit, and the terms in the DE are voltages.

We briefly describe the derivation from physics. For a capacitor,

$$q(t) = Cv(t),$$

where C is the capacitance. So

$$v_{capacitor}(t) = \frac{1}{C}q(t).$$

The voltage across an inductor is equal to

$$v_{inductor}(t) = L\frac{d^2q}{dt^2},$$

where L is the inductance. The voltage across a resistor is equal to

$$v_{resistor}(t) = R\frac{dq}{dt},$$

where L is the resistance (this is Ohm's Law).

From Kirchhoff's Law, these voltages all sum to zero, so we obtain

$$Lq'' + Rq' + \frac{1}{C}q = 0.$$

If an external voltage $E(t)$ source is added to the system, then the sum will not be zero, but will need to be $E(t)$. So we obtain

$$Lq'' + Rq' + \frac{1}{C}q = E(t).$$

Note that if voltage in this DE is measured in volts, then $q(t)$ is measured in Coulombs, L are Henries, R is in Ohms, C is in Farads, and $q'(t)$ is amperes (amps).

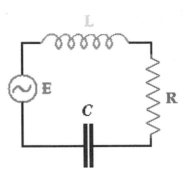

Figure 3.11: **An LRC-circuit forced by E(t)**

Example 3.28 *An LRC-circuit is attached to a 9 volt battery where $L = 1$ Henry, $R = 2$ Ohms, and C is $\frac{1}{3}$ Farads. Find the steady state for $q(t)$.*

Solution: Using undetermined coefficients, we set $q(t) = A$ and plug this into the DE

$$q'' + 2q' + 3q = 9.$$

We obtain $q(t) = 3$, so the charge limits to 3 coulombs. Note that this would imply that the capacitor would eventually have a voltage difference of 9 volts across it.   □

Example 3.29 *An LRC-circuit is attached to a 12 volt battery is turned on and off periodically so that $E(t) = 6\sin t + 6$. If the circuit has $L = 1$ Henry, $R = 2$ Ohms, and C is $\frac{1}{3}$ Farads. Find the steady state for $q(t)$.*

Solution: Using undetermined coefficients, we set $q(t) = A\sin t + B\cos t + C$. From plugging to the DE:

$$-A\sin t + -B\cos t - 2B\sin t + 2A\cos t + 3A\sin t + 3B\cos t + 3C = 6\sin t + 6$$

We obtain $2A - 2B = 1$ and $2A + 2B = 0$ and $C = 2$ so $A = \frac{1}{4}$, $B = -\frac{1}{4}$ and $C = 2$. Thus, $q(t) = \frac{1}{4}\sin t - \frac{1}{4}\cos t + 2$.   □

As one can see, LRC circuits behave in exactly the same manner as spring mass systems. Hence, just as spring mass systems, LRC circuits can be characterized as overdamped, underdamped, or critically damped, and can exhibit resonance if the resistance is low enough.

The reader should be aware that in many applications, it is $q'(t)$ which is of interest (current) and not $q(t)$ (charge) itself. In this case, the goal is to maximize the amplitude of the current $(q'(t))$. It turns out that amplitude of sinusoidal forcing always exceeds the steady state by forcing with a constant voltage, since the steady state for constant forcing has $q'(t) = 0$ (current limits to zero).

Resonance in Sinusoidally Driven RLC-circuits

The sinusoidally forced spring mass system

$$Lq'' + Rq' + \frac{1}{C}q = V_0 \cos(\omega t - \phi) \tag{3.21}$$

has a steady state solution with maximal amplitude for current $q'(t)$ when

$$\omega = \frac{1}{\sqrt{LC}}.$$

This is called **the resonance frequency for current.** It optimizes current through the circuit.

The frequency that yields the maximal amplitude charge on the capacitor for the steady state only occurs when the system is lightly damped ($2\frac{L}{C} - R^2 > 0$) and has frequency:

$$\omega = \frac{\sqrt{2\frac{L}{C} - R^2}}{\sqrt{2}L} \tag{3.22}$$

This is called **the resonance frequency for charge on the capacitor.** It produces the largest voltage difference across the capacitor. (If $2\frac{L}{C} - R^2 < 0$ then constant forcing produces the largest voltage difference across the capacitor.)

Note that the optimal frequency it depends upon what one is trying to maximize. Again, usually it is an optimal current that is of most interest in many RLC-circuits applications.

Exercises

1. An RLC circuit is attached to a 9 volt battery (constant). The steady state has a charge of 2 Coulombs on the capacitor. Find the capacitance C for the capacitor.

2. An LRC circuit has inductor with inductance 9 Henries, resistor with resistance 2 Ohms, and capacitor with capacitance 1 Farads. Is the circuit overdamped, or underdamped? What would a typical solution's graph look like?

3. An LRC circuit has inductor with inductance 12 Henries, resistor with resistance 4 Ohms, and capacitor with capacitance 2 Farads. If the initial charge on the capacitor is 5 coulombs, with initial current of 2 amps (with charge on the capacitor increasing), find the charge on the capacitor at any time t.

4. An LRC circuit has inductor with inductance 1 Henry, resistor with resistance 2 Ohms, and capacitor with capacitance 3 Farads. If the initial charge on the capacitor is 1 coulomb, with initial current of 3 amps (with charge on the capacitor decreasing), find the charge on the capacitor at any time t.

5. An LRC circuit has inductor with inductance 3 Henries, resistor with resistance 1 Ohm, and capacitor with capacitance $\frac{1}{10}$ Farads. It is attached to an 10 volt battery. Find the steady state solution.

6. LRC circuits come in 3 varieties, Find conditions on L, R, and C so that the circuit is (a) overdamped (b) critically damped (c) underdamped (d) lightly underdamped. Find the resonance frequency (for optimal amplitude for charge) for a sinusoidally driven lightly underdamped LRC circuit.

7. For RLC-circuits, one normally seeks to maximize the amplitude steady state's current which is $q'(t)$. For fixed input voltage functions of the form $V_0 \cos(\omega t - \phi)$, use undetermined coefficients to show that this happens when $\omega = \frac{1}{\sqrt{LC}}$ for all LRC circuits. Note that this is true for overdamped, underdamped and critically damped LRC-circuits.

8. An unforced RLC circuit experiences multiple changes in the direction of the current for all $t > 0$ for certain initial conditions. What can be said about $R^2 - 4\frac{L}{C}$? Explain.

9. An unforced RLC circuit experiences only one change in the direction of the current for all $t > 0$ for certain initial conditions. What can be said about $R^2 - 4\frac{L}{C}$? Explain.

10. An AC voltage source given by $5\cos(2t)$ volts is attached to an LRC circuit with inductance 1 Henry, resistor with resistance 1 Ohms, and capacitor with capacitance 1 Farads. If there is no charge and no current through the capacitor, find the charge on the capacitor at any time t.

11. An AC voltage source given by $9\cos(\omega t)$ volts is attached to an LRC circuit with inductance 2 Henries, resistor with resistance 1 Ohms, and capacitor with capacitance 1 Farads.

    (a) Find the amplitudes of both $q(t)$ and $q'(t)$ for the steady state solution for $\omega = \frac{1}{\sqrt{LC}}$.

    (b) Find the amplitudes of both $q(t)$ and $q'(t)$ for the steady state solution for $\omega = \frac{\sqrt{2\frac{L}{C}-R^2}}{\sqrt{2}L}$.

12. An LRC circuit has inductor with inductance 4 Henries, resistor with resistance 2 Ohms, and capacitor with capacitance 1 Farads. It is forced with $F(t) = 12\cos(\omega t)$

    (a) Find the amplitudes of both $q(t)$ and $q'(t)$ for the steady state solution for $\omega = \frac{1}{\sqrt{LC}}$.

    (b) Find the amplitudes of both $q(t)$ and $q'(t)$ for the steady state solution for $\omega = \frac{\sqrt{2\frac{L}{C}-R^2}}{\sqrt{2}L}$.

13. An AC voltage source given by $V_0\cos(\omega t)$ volts is attached to an LRC circuit with inductance L Henry, resistor with resistance R Ohms, and capacitor with capacitance C Farads. Find the frequency ω that yields the largest amplitude for q'' over all values of ω.

3.8   Wronskians and Variation of Parameters

In this section, we will need the concept of the Wronskian defined below:

The Wronskian

<u>Wronskian of 2 functions</u>

Define the Wronskian of two differentiable functions $f(t)$ and $g(t)$ to be

$$W[f, g](t) = \begin{vmatrix} f(t) & g(t) \\ f'(t) & g'(t) \end{vmatrix} = f(t)g'(t) - g(t)f'(t)$$

Example 3.30 *Compute the Wronksian of $f(t) = t^2$ and $g(t) = \sin(6t)$*

Solution:

$$W[f, g](t) = \begin{vmatrix} f(t) & g(t) \\ f'(t) & g'(t) \end{vmatrix} = f(t)g'(t) - g(t)f'(t),$$

so

$$W[t^2, \sin(6t)] = \begin{vmatrix} t^2 & \sin(6t) \\ 2t & 6\cos(6t) \end{vmatrix}$$

$$= 6t^2 \cos(6t) - 2t \sin(6t). \quad \square$$

Example 3.31 *Compute the Wronksian of $f(t) = \sqrt{t}$ and $g(t) = e^{2t}$*

Solution:

$$W[f, g](t) = \begin{vmatrix} f(t) & g(t) \\ f'(t) & g'(t) \end{vmatrix} = f(t)g'(t) - g(t)f'(t),$$

so

$$W[\sqrt{t}, e^{2t}] = \begin{vmatrix} \sqrt{t} & e^{2t} \\ \frac{1}{2\sqrt{t}} & 2e^{2t} \end{vmatrix}$$

$$= 2\sqrt{t}e^{2t} - \frac{1}{2\sqrt{t}}e^{2t}. \quad \square$$

Notice that the Wronskian of two functions is again a new function whose domain depends upon the domains of f and g and their derivatives as illustrated in the previous example. There are some properties of the Wronskian, which are straightforward to verify.

1. $W[0, g(t)] = 0$

2. $W[f(t), f(t)] = 0$

3. $W[1, g(t)] = g'(t)$

4. $W[f(t), g(t) + h(t)] = W[f(t), g(t)] + W[f(t), h(t)]$

5. $W'[f(t), g(t)] = fg'' - f''g = W[f, g'] + W[f', g]$

6. For any constant c, $W[f(t), cg(t)] = cW[f(t), g(t)] = W[cf(t), g(t)]$

7. $W[f(t), g(t)] = -W[g(t), f(t)]$

The last property tells us that the order of the functions in the Wronskian is important.

Variation of Parameters

Suppose that

$$y''(t) + a(t)y'(t) + b(t)y(t) = F(t)$$

is a second order linear ODE and that $c_1 y_1(t) + c_2 y_2(t)$ is a general solution to the associated homogeneous DE. Then

---

Variation of Parameters

Suppose that

$$y''(t) + a(t)y'(t) + b(t)y(t) = F(t)$$

is a second order linear ODE and that $c_1 y_1(t) + c_2 y_2(t)$ is a general solution to the associated homogeneous DE.
Then

$$y_P(t) = v_1(t)y_1(t) + v_2(t)y_2(t)$$

is a particular solution to the nonhomogeneous DE, where

$$v_1(t) = -\int \frac{F(t)y_2(t)}{W[y_1, y_2]}\, dt$$

$$v_2(t) = \int \frac{F(t)y_1(t)}{W[y_1, y_2]}\, dt$$

(As usual, the antiderivatives in the formulas for v_1, v_2 denote any one antiderivative.)

Example 3.32 *Find a solution to the nonhomogeneous DE*

$$y'' + 4y = \sec(2t)$$

Solution: Note that the general solution to the homogeneous DE is

$$y_{homog}(t) = c_1 \cos(2t) + c_2 \sin(2t),$$

so $y_1 = \cos(2t)$ and $y_2 = \sin(2t)$.

Next, note that

$$W[y_1, y_2] = W[\cos(2t), \sin(2t)] = \begin{vmatrix} \cos(2t) & \sin(2t) \\ -2\sin(2t) & 2\cos(2t) \end{vmatrix}$$

$$= 2\cos^2(2t) + 2\sin^2(2t) = 2.$$

So

$$v_1(t) = -\int \frac{\sec(2t)\sin(2t)}{2} \, dt = -\int \frac{\tan(2t)}{2} \, dt = \frac{1}{4}\ln|\sin(2t)|$$

and

$$v_2(t) = \int \frac{\sec(2t)\cos(2t)}{2} \, dt = \int \frac{1}{2} \, dt = \frac{1}{2}t$$

Thus $y_P(t) = v_1 y_1 + v_2 y_2 = \frac{1}{4}\ln|\sin(2t)|\cos(2t) + \frac{1}{2}t\sin(2t)$ solves the non-homogenous DE.

Note that the general solution to the DE (by the Superposition Principle) is $y(t) = \frac{1}{4}\ln|\sin(2t)|\cos(2t) + \frac{1}{2}t\sin(2t) + c_1\cos(2t) + c_2\sin(2t)$.   □

The next example illustrates that this method works even when the coefficients are not all constant (unlike the method of undetermined coefficients). Note in the example that the DE needs to be put into standard form before we can use the formula.

Example 3.33 *Find a solution to the nonhomogeneous DE*

$$t^2 y'' - 2y = \frac{3}{t}, \quad t \neq 0$$

given that $y(t) = c_1 t^2 + c_2 \frac{1}{t}$ *solves*

$$t^2 y'' - 2y = 0$$

Solution: Note that the general solution to the homogeneous DE is

$$y_{homog}(t) = c_1 t^2 + c_2 \frac{1}{t},$$

so $y_1 = t^2$ and $y_2 = \frac{1}{t}$. Further note that in standard form, the DE becomes

$$y'' - \frac{2}{t^2} y = \frac{3}{t^3}, \quad t \neq 0$$

Next, note that

$$W[y_1, y_2] = W[t^2, \frac{1}{t}] = \begin{vmatrix} t^2 & \frac{1}{t} \\ 2t & -\frac{1}{t^2} \end{vmatrix}$$

$$= -1 - 2 = -3.$$

So

$$v_1(t) = -\int \frac{\frac{3}{t^4}}{-3} \, dt = \int \frac{1}{t^4} \, dt = -\frac{1}{3} \frac{1}{t^3}$$

and

$$v_2(t) = \int \frac{\frac{3}{t^3} t^2}{-3} \, dt = -\int \frac{1}{t} \, dt = -\ln|t|$$

Thus $y_P(t) = v_1 y_1 + v_2 y_2 = -\frac{1}{3t^3} t^2 - \frac{\ln|t|}{t}$ solves the nonhomogenous DE. $\square$

As one might guess, the previous examples were chosen very carefully so that the antiderivatives could be computed in closed form. At first, it might seem that this method allows us only to solve a small set of problems, in particular, problems where the antiderivatives in the formula are computable. However, this method becomes extremely powerful and versatile if we recall that the antiderivatives of $G(t)$ are simply obtained by $\int_{t_0}^t G(w) \, dw$, where (t_0, t) is in the domain of G. Hence, the variation of parameters method allows us to obtain a particular solution even when the antiderivatives do not "work out nicely". The tradeoff is that one may need to approximate a definite integral to evaluate a solution as in the next example.

Example 3.34 *Find the solution to the nonhomogeneous ODE/IVP*

$$y'' + 4y = \ln(t + 1), \quad y(0) = 0, \quad y'(0) = 1$$

and use it to approximate $y(2)$

Solution: As in the previous example, the general solution to the homogeneous DE is

$$y_{homog}(t) = c_1 \cos(2t) + c_2 \sin(2t),$$

so $y_1 = \cos(2t)$ and $y_2 = \sin(2t)$ and

$$W[y_1, y_2] = W[\cos(2t), \sin(2t)] = \begin{vmatrix} \cos(2t) & \sin(2t) \\ -2\sin(2t) & 2\cos(2t) \end{vmatrix}$$

$$= 2\cos^2(2t) + 2\sin^2(2t) = 2.$$

So

$$v_1(t) = -\int \frac{\ln(t)\sin(2t)}{2}\, dt = -\frac{1}{2}\int_0^t \ln(w+1)\sin(2w)\, dw$$

and

$$v_2(t) = \int \frac{\ln(t)\cos(2t)}{2}\, dt = \frac{1}{2}\int_0^t \ln(w+1)\cos(2w)\, dw$$

(we take $t_0 = 0$).

So the general solution to the nonhomogeneous DE is

$$y(t) = c_1 \cos(2t) + c_2 \sin(2t)$$

$$-\cos(2t)\left(\frac{1}{2}\int_0^t \ln(w+1)\sin(2w)\, dw\right) + \sin(2t)\left(\frac{1}{2}\int_0^t \ln(w+1)\cos(2w)\, dw\right)$$

Since $y(0) = 0$ we have

$$0 = c_1 - \left(\frac{1}{2}\int_0^0 \ln(w+1)\sin(2w)\, dw\right)$$

so $c_1 = 0$

Using the product rule and the second fundamental theorem of calculus,

$$y'(t) = 2c_2 \cos(2t) - \cos(2t)\left(\frac{1}{2}\ln(t+1)\sin(2t)\right)$$

$$+2\sin(2t)\left(\frac{1}{2}\int_0^t \ln(w+1)\sin(2w)\, dw\right) + 2\cos(2t)\left(\frac{1}{2}\int_0^t \ln(w+1)\cos(2w)\, dw\right)$$

$$+\sin(2t)\left(\frac{1}{2}\ln(t+1)\cos(2t)\right)$$

Plugging in $t = 0$ yields $1 = 2c_2$ so $c_2 = \frac{1}{2}$.

So the particular solution we seek is

$$y(t) = \frac{1}{2}\sin(2t) - \cos(2t)\left(\frac{1}{2}\int_0^t \ln(w+1)\sin(2w)\ dw\right) + \sin(2t)\left(\frac{1}{2}\int_0^t \ln(w+1)\cos(2w)\ dw\right)$$

$$y(2) = \frac{1}{2}\sin(4) - \cos(4)\left(\frac{1}{2}\int_0^2 \ln(w+1)\sin(2w)\ dw\right)$$

$$+ \sin(4)\left(\frac{1}{2}\int_0^2 \ln(w+1)\cos(2w)\ dw\right) \approx .0014974951$$

Above, we obtained an approximation of $y(2)$ by approximating the definite integrals (you could approximate the integrals Simpson's method, the Trapezoidal method., Upper/Lower sums, or simply a calculator that can approximate definite integrals).

Exercises

Use the variation of parameters method to find a particular solution to the DE

1. $y'' + 3y' + 2y = 4e^t$

2. $y'' + 3y' + 2y = t$

3. $y'' + y = \tan t$

4. $y'' + y = \csc t$

5. $y'' + 4y = \sin(2t)\cos(2t)$

Use the variation of parameters method to find a general solution to the DE

6. $y'' + 9y = \cot(3t)$

7. $y'' + y = \csc t$

8. $y'' + 4y = \sin(2t)\cos(2t)$

9. $t^2 y'' - 6y = t^4$ given that $y(t) = c_1 t^3 + c_2 \frac{1}{t^2}$ solve the homogeneous DE. (Hint: Put the DE in standard form first!)

Use the variation of parameters method

10. $y'' + 4y' + 3y = 5$, $y(1) = 1$, $y'(1) = 2$

11. $y'' + 9y = \cos(2t)$, $y(1) = 0$, $y'(1) = 2$

12. $y'' + 4y = t$, $y(0) = 1$, $y'(0) = -1$

Use the variation of parameters method to approximate $y(2)$ for the IVP below. The resulting integrals will need to be approximated.

13. $y'' + y = \frac{\sin t}{t}$, $y(0) = 0$, $y'(0) = 1$

14. $y'' + y = \frac{1}{t+1}$, $y(0) = 0$, $y'(0) = 1$

3.9 Linear Independence, General Solutions, and the Wronskian

Linear Dependence and Independence

We start with a definition:

Two functions $f(t)$ and $g(t)$ are called **linearly dependent on an interval** I if there exists **non-zero** constants c_1 and c_2 so that

$$c_1 f(t) + c_2 g(t) = 0$$

for all t in I. (Otherwise, the functions are called **linearly independent on** I). Note that it is easy to satisfy $c_1 f(t) + c_2 g(t) = 0$ by taking $c_1 = c_2 = 0$, hence the condition that the constants be non-zero is important.

Note that if $f(t)$ and $g(t)$ are linearly dependent on an interval I, then since $c_1 \neq 0$, we can write $f(t) = -\frac{c_2}{c_1} g(t)$ and so f is a constant multiple of g on I.

Example 3.35 *Show that $f(t) = t$ and $g(t) = t^2$ are linearly independent on $(-\infty, \infty)$.*

Solution: We will show that these functions cannot satisfy

$$c_1 f(t) + c_2 g(t) = 0$$

for fixed nonzero constants on all of $(-\infty, \infty)$. To see this suppose $c_1 t + c_2 t^2 = 0$ for all t in $(-\infty, \infty)$. Since $c_1 t + c_2 t^2 = 0$ must be true for all t, it must be true for $t = 1$ (so we obtain $c_1 + c_2 = 0$, so $c_1 = -c_2$), and it must be true for $t = -1$ (so we obtain $-c_1 + c_2 = 0$, so $c_1 = c_2$). These two conditions imply that $c_1 = c_2 = 0$. So there cannot exist non-zero constants so that $c_1 t + c_2 t^2 = 0$ for all t in $(-\infty, \infty)$. This means that $f(t) = t$ and $g(t) = t^2$ are linearly independent on $(-\infty, \infty)$. $\qquad\square$

Theorem 3.36 *If $f(t)$ and $g(t)$ are linearly dependent and both differentiable on an interval I, then $W[f, g] = 0$ on I.*

Proof: Suppose that there exists non-zero constants c_1 and c_2 so that

$$c_1 f(t) + c_2 g(t) = 0$$

for all t in I. Then $f(t) = -\frac{c_2}{c_1} g(t)$ on I. Then, by properties of the Wronskian proven in the previous section, for any t in I we have

$$W[f(t), g(t)] = W[-\frac{c_2}{c_1} g(t), g(t)] = -\frac{c_2}{c_1} W[g(t), g(t)] = 0. \quad \square$$

The above theorem implies the result below:

Theorem 3.37 *If $W[f, g](t_0) \neq 0$ for some value t_0 in I then $f(t)$ and $g(t)$ are linearly independent on the interval I.*

Example 3.38 *Use the Theorem 3.37 to show that $f(t) = t$ and $g(t) = t^2$ are linearly independent on any interval I.*

Solution: The Wronskian of $f(t) = t$ and $g(t) = t^2$ is

$$W[t, t^2] = \begin{vmatrix} t & t^2 \\ 1 & 2t \end{vmatrix}$$

$$= 2t^2 - t^2 = t^2.$$

This is only zero for $t = 0$, so any interval I will contain a value $t_0 \neq 0$ so that $W[t, t^2](t_0) \neq 0$. So $f(t) = t$ and $g(t) = t^2$ are linearly independent on I. $\square$

One should be very careful when trying to use the Wronskian to deduce linear dependence of two functions as the next example illustrates.

Example 3.39 *Let $f(t) = t^3$ and $g(t) = |t^3|$*
   (a) Show that $W[f, g](t) = 0$ for all t.
   (b) Show that $f(t)$ and $g(t)$ are linearly dependent on $(0, \infty)$.
   (c) Show that $f(t)$ and $g(t)$ are linearly independent on $(-\infty, \infty)$.

Solution: The Wronskian of $f(t) = t^3$ and $g(t) = |t^3|$ for $t \geq 0$ is clearly zero since $g(t) = t^3$ for $t \geq 0$. For $t \leq 0$, we have $g(t) = -t^3$, and so, $W[f, g](t) = 0$ for all t.

To see that these are linearly dependent on $(0, \infty)$ follows from the fact that for all $t > 0$ $f(t) = g(t) = t^3$. In other words,

$$c_1 f(t) + c_2 g(t) = 0$$

can be solved by $c_1 = 1$ and $c_2 = -1$.

To see that these are not linearly dependent on the interval $(-\infty, \infty)$. We solve

$$c_1 f(t) + c_2 g(t) = 0.$$

When $t = 1$ we would have

$$c_1(1)^3 + c_2(1)^3 = 0 \quad \text{or} \quad c_1 + c_2 = 0$$

and for $t = -1$ we would have

$$c_1(-1)^3 + c_2|(-1)^3| = 0 \quad \text{or} \quad -c_1 + c_2 = 0.$$

Taking these together, $c_1 = 0$ and $c_2 = 0$ so there is no nonzero choice of c_1 and c_2 that will solve $c_1 f(t) + c_2 g(t) = 0$ for all t in $(-\infty, \infty)$. $\qquad \square$

The previous example also illustrates that it is necessary to state the interval of consideration when discussing linear independence and dependence. The next theorem relates the Wronskian to solving initial value problems.

Theorem 3.40 *Suppose that $y_1(t)$ and $y_2(t)$ are differentiable at t_0 and*

$$y(t) = c_1 y_1(t) + c_2 y_2(t)$$

then one can solve for c_1 and c_2 uniquely to satisfy $y(t_0) = A$ and $y'(t_0) = B$ for all possible values of A and B if, and only if

$$W[y_1, y_2](t_0) \neq 0.$$

Proof: For any initial conditions $y(t_0) = A$ and $y'(t_0) = B$, being able to solve for c_1 and c_2 amounts to being able to solve the system:

$$c_1 y_1(t_0) + c_2 y_2(t_0) = A$$

$$c_1 y_1'(t_0) + c_2 y_2'(t_0) = B$$

can always be solved for c_1 and c_2.

By elimination, we see that

$$c_1 = \frac{A y_2'(t_0) - B y_2(t_0)}{y_1(t_0) y_2'(t_0) - y_2(t_0) y_1'(t_0)}$$

$$c_2 = \frac{B y_1(t_0) - A y_1'(t_0)}{y_1(t_0) y_2'(t_0) - y_2(t_0) y_1'(t_0)}$$

Since the denominator of the above terms is $W[y_1, y_2](t_0)$, we see that $W[y_1, y_2](t_0) \neq 0$ is necessary and sufficient to solve for c_1 and c_2 for all possible A and B. $\quad \square$

The previous Theorem tells us that in order to solve an arbitrary initial value problem $y(t_0) = A$ and $y'(t_0) = B$ where $y(t) = c_1 y_1(t) + c_2 y_2(t)$, the Wronskian of y_1 and y_2 at t_0 must be non-zero.

Some Additional Theorems

In this section, we state several additional Theorems without proof. The proofs of these theorems are beyond the scope of a basic treatment of this subject.

Theorem 3.41 (Existence and Uniqueness for Second Order Linear ODE) *Suppose that $p(t)$, $q(t)$ and $f(t)$ are continuous on the interval I with t_0 in I. Then the ODE/IVP*

$$\frac{d^2 y}{dt^2} + p(t)\frac{dy}{dt} + q(t)y(t) = f(t), \quad y(t_0) = A, \quad y'(t_0) = B$$

has a unique solution that exists on all of I.

Theorem 3.42 *Suppose that $y(t) = c_1 y_1(t) + c_2 y_2(t)$, solves*

$$\frac{d^2 y}{dt^2} + p(t)\frac{dy}{dt} + q(t)y(t) = f(t), \quad y(t_0) = A, \quad y'(t_0) = B$$

on the interval I with t_0 in I. If $W[y_1, y_2](t_0) \neq 0$ then $W[y_1, y_2](t) \neq 0$ for all t in I and $y(t) = c_1 y_1(t) + c_2 y_2(t)$ is a general solution to the ODE on I.

This Theorem is a Corollary of Abel's Theorem, which we do not state here.

Exercises

Use the Wronskian to show that the following are linearly independent on $(-\infty, \infty)$

1. $f(t) = \cos t; \quad g(t) = \sin t$

2. $f(t) = e^t \quad g(t) = e^{-t}$

3. $f(t) = 1 \quad g(t) = t$

4. Let $f(t) = t \quad g(t) = t^2$.

    (a) Show that $W[f, g](0) = 0$

    (b) Are the functions linearly independent on $(-\infty, \infty)$?

5. Let $f(t) = t \quad g(t) = |t|$.

    (a) Use the definition (not the Wronskian) to show that the functions are linearly dependent on $(0, \infty)$

    (b) Use the definition (not the Wronskian) to show that the functions are linearly independent on $(-\infty, \infty)$

    (c) Why can't we use the Wronskian at all in part (b)?

Chapter 4

nth Order Linear Equations

4.1 nth Order Linear ODE –General Results

Recall that an nth ODE is linear if it can be written in the form:

$$a_n(t)y^{(n)} + a_{n-1}(t)y^{(n-1)} + ... + a_2(t)y'' + a_1(t)y' + a_0(t)y = f(t),$$

As before, if $f(t)$ is zero, we call the DE homogeneous. Possibly the most important aspect of linear differential equations is the superposition principle which we have already seen in the second order case. We state the general result below:

Superposition Principle

Suppose $y_1(t)$ solves the linear DE

$$a_n(t)y^{(n)} + a_{n-1}(t)y^{(n-1)} + ... + a_2(t)y'' + a_1(t)y' + a_0(t)y = f(t) \qquad (4.1)$$

and that $y_2(t)$ solves the linear DE

$$a_n(t)y^{(n)} + a_{n-1}(t)y^{(n-1)} + ... + a_2(t)y'' + a_1(t)y' + a_0(t)y = g(t) \qquad (4.2)$$

Then $Y(t) = c_1 y_1(t) + c_2 y_2(t)$ solves

$$a_n(t)y^{(n)} + a_{n-1}(t)y^{(n-1)} + ... + a_2(t)y'' + a_1(t)y' + a_0(t)y = c_1 f(t) + c_2 g(t) \quad (4.3)$$

The proof of this result is similar to the one provided for the second order case. This result has two immediate applications:

1. If $y_1(t)$ and $y_2(t)$ are solutions of a linear homogeneous ODE, then any linear combination of them will also solve the same ODE.

2. If $y_P(t)$ solves the nonhomogeneous linear ODE:

$$a_n(t)y^{(n)} + a_{n-1}(t)y^{(n-1)} + ... + a_2(t)y'' + a_1(t)y' + a_0(t)y = f(t),$$

and $y_{homog}(t)$ solves the associated homogenous ODE, then the sum $y_{homog}(t) + y_P(t)$ also solves the nonhomogeneous ODE.

The functions $\{f_1(t), f_2(t), ..., f_n(t)\}$ are called **linearly dependent on an interval** I if there exists constants $c_1, c_2, ..., c_n$, of which at least one is not zero, with

$$c_1 f_1(t) + c_2 f_2(t) + ... + c_n f_n(t) = 0$$

for all t in I. (Otherwise, the functions are called **linearly independent on** I). A function $g(t)$ is a **linear combination** of the functions $\{f_1(t), f_2(t), ..., f_n(t)\}$ if there exists constants $c_1, c_2, ..., c_n$, so that

$$g(t) = c_1 f_1(t) + c_2 f_2(t) + ... + c_n f_n(t)$$

for all t in I. If $\{f_1(t), f_2(t), ..., f_n(t)\}$ is a linearly independent list, then no function in the list can be a linear combination of the other functions.

Example 4.1 *Are* $\{e^t, \sin t, t\}$ *linearly independent or dependent on* $(-\infty, \infty)$*?*

Solution: We suppose that

$$c_1 e^t + c_2 \sin t + c_3 t = 0$$

for all t. We will plug in several values of t into the above statement.

For $t = 0$, we see that

$$c_1 e^0 + c_2 \sin 0 + c_3 0 = 0,$$

so $c_1 = 0$. For $t = \pi$, we see that $c_2 \sin \pi + c_3 \pi = 0$, so $c_3 = 0$. Lastly, plugging in $t = \frac{\pi}{2}$ we get $c_2 = 0$. This implies that the only way to solve

$$c_1 e^t + c_2 \sin t + c_3 t = 0$$

for all t is to have $c_1 = c_2 = c_3 = 0$, so the list is linearly independent on $(-\infty, \infty)$. □

Example 4.2 *Are* $\{\sin^2 t, \cos^2 t, 1\}$ *linearly independent or dependent on* $(-\infty, \infty)$?

Solution: Since

$$c_1 \sin^2 t + c_2 \cos^2 t + c_3 1 = 0$$

is true for all t when we take $c_1 = 1, c_2 = 1$, and $c_3 = -1$ (since $\sin^2 t + \cos^2 t = 1$), we see that $\{\sin^2 t, \cos^2 t, 1\}$ are linearly dependent on $(-\infty, \infty)$. $\quad\square$

An nth order linear ODE must have n free constants in order solve an arbitrary any initial value problem. So a general solution to an nth order linear must be a linear combination of n linearly independent functions. We define a tool to help us deduce whether functions are linearly independent or not on an interval I.

Wronskian of n functions

Define the Wronskian of a list of n functions $\{f_1(t), f_2(t), ..., f_n(t)\}$ to be

$$W\left[f_1(t), f_2(t), ..., f_n(t)\right](t) = \begin{vmatrix} f_1(t) & f_2(t) & \cdots & f_n(t) \\ f_1'(t) & f_2'(t) & \cdots & f_n'(t) \\ \vdots & \vdots & \cdots & \vdots \\ f_1^{(n-1)}(t) & f_2^{(n-1)}(t) & \cdots & f_n^{(n-1)}(t) \end{vmatrix}$$

We note that for two functions $f_1(t)$ and $f_2(t)$,

$$W[f_1(t), f_2(t)] = f_1(t)f_2'(t) - f_1'(t)f_2(t).$$

For three functions,

$$W[f_1(t), f_2(t), f_3(t)] = f_1(t)f_2'(t)f_3''(t) + f_2(t)f_3'(t)f_1''(t) + f_3(t)f_1'(t)f_2''(t)$$

$$- f_1(t)f_3'(t)f_2''(t) - f_2(t)f_1'(t)f_3''(t) - f_3(t)f_2'(t)f_1''(t).$$

Higher order Wronskians can be computed using techniques for computing determinants from linear algebra. The next result has a proof that comes from a basic understanding of linear algebra.

Theorem 4.3 (Wronskians of Linearly Dependent Functions) *Suppose that* $f_1(t), ..., f_n(t)$ *are linearly dependent on* I *and have* $n-1$ *derivatives. Then* $W[f_1(t), ..., f_n(t)] = 0$ *on* I.

Proof: Suppose that $c_1 f_1(t) + ... + c_n f_n(t) = 0$, where not all of the c_i are zero. Then, by differentiating, it is also true that

$$c_1 f_1'(t) + ... + c_n f_n'(t) = 0,$$

$$c_1 f_1''(t) + ... + c_n f_n''(t) = 0,$$

and, ...

$$c_1 f_1^{(n-1)}(t) + ... + c_n f_n^{(n-1)}(t) = 0.$$

This gives a homogeneous linear system of n equations in n variables $c_1, ..., c_n$. Since there is a non-trivial solution to this homogeneous linear system,

$$W\left[f_1(t), f_2(t), ..., f_n(t)\right](t) = \begin{vmatrix} f_1(t) & f_2(t) & \cdots & f_n(t) \\ f_1'(t) & f_2'(t) & \cdots & f_n'(t) \\ \vdots & \vdots & \cdots & \vdots \\ f_1^{(n-1)}(t) & f_2^{(n-1)}(t) & \cdots & f_n^{(n-1)}(t) \end{vmatrix} = 0$$

for all t in I. $\square$

Note that the theorem does NOT say that if the Wronskian is zero, we are not guaranteed linear dependence. This Theorem only says if the functions are linearly independent, then the Wronskian is zero.

There are some properties of the Wronskian.

1. $W[0, f_2(t), f_3(t), ..., f_n(t)] = 0$

2. $W[f_1(t), f_1(t), f_3(t), ..., f_n(t)] = 0$

3. $W[1, f_2(t), f_3(t), ..., f_n(t)] = W[f_2'(t), f_3'(t), ..., f_n'(t)]$

4. For any constant c,

$$W[c f_1(t), f_2(t), f_3(t), ..., f_n(t)] = c W[f_1(t), f_2(t), f_3(t), ..., f_n(t)]$$

5.

$$W[f_1(t), f_2(t), ..., f_i(t), ..., f_j(t), ..., f_n(t)]$$
$$= -W[f_1(t), f_2(t), ..., f_j(t), ..., f_i(t), ..., f_n(t)]$$

Example 4.4 *Consider*

$$y^{(3)} + y'' + y' + y = 3t + 3$$

Suppose we know that $y_{homog}(t) = c_1 e^{-t} + c_2 \cos t + c_3 \sin t$ is the general solution to the associated homogeneous DE and that $y_P(t) = 3t$ solves the nonhomogeneous ODE. Then $Y(t) = c_1 e^{-t} + c_2 \cos t + c_3 \sin t + 3t + 3$ will also solve the nonhomogeneous ODE.

$\square$

4.2 Existence/Uniqueness Theorem

We note that in order to specify an initial value problem at some specified value t_0, one must specify n conditions, namely $y(t_0), y'(t_0), ..., y^{(n-1)}(t_0)$. Once one has posed such an IVP we again have the Existence and Uniqueness Theorem for Linear ODE:

Theorem 4.5 (Existence and Uniqueness for Linear ODE) *Suppose that $a_{n-1}(t), a_{n-2}(t), ..., a_1(t), a_0(t), f(t)$ are all continuous on an open interval I and suppose that t_0 is in I.*

Then the ODE/IVP:

$$y^{(n)} + a_{n-1}(t)y^{(n-1)} + ... + a_2(t)y'' + a_1(t)y' + a_0(t)y = f(t) \qquad (4.4)$$

with any specified values for $y(t_0), y'(t_0), ..., y^{(n-1)}(t_0)$ has a unique solution that exists on (at least) all of I.

Example 4.6 *Find the largest interval (a, b) for which the Existence/Uniqueness Theorem guarantees as a unique solution to the ODE/IVP.*

$$y^{(3)} + (\sin t)t^2 y' + (\tan t)\, y = \frac{1}{t - 10}, \ \ y(1) = 1, \ \ y'(1) = 2, \ \ y''(1) = 3$$

Solution: Since $t_0 = 1$ and since each of the functions: $t^2 \sin t, \tan t$, and $\frac{1}{t-10}$ are continuous at $t_0 = 1$, do know that there exists a unique solution on some interval. The largest such interval that contains $t_0 = 1$ for which all of these functions are continuous is $(-\frac{\pi}{2}, \frac{\pi}{2})$. So, we know a unique solution exists to this IVP and can guarantee its existence on $(-\frac{\pi}{2}, \frac{\pi}{2})$.   □

The next example shows that we may not have unique solutions to a nonlinear ODE/IVP.

Example 4.7 *Show that* $y(t) = (\frac{2}{3}t)^{\frac{3}{2}}$ *satisfies the ODE/IVP*

$$y' - y^{\frac{1}{3}} = 0, \ y(0) = 0.$$

Find a different solution to the same IVP.

Solution: First note that the DE is not linear, so the Existence/Uniqueness Theorem does not apply. Now for $y(t) = (\frac{2}{3}t)^{\frac{3}{2}}$, clearly $y(0) = 0$. Next, we see that $y'(t) = (\frac{2}{3}t)^{\frac{1}{2}}$ so

$$y' - y^{\frac{1}{3}} = (\frac{2}{3}t)^{\frac{1}{2}} - \left((\frac{2}{3}t)^{\frac{3}{2}} \right)^{\frac{1}{3}} = 0.$$

A second solution is $Y(t) = 0$ which also solves the ODE/IVP, so the ODE/IVP does not have a unique solution.   □

The next example shows some of the power of this remarkable Theorem.

Example 4.8 *Use the Existence/Uniqueness Theorem to show that* $y(t) = \tan t$ *cannot solve the ODE*

$$y'' + t^2 y' + y = \frac{1}{t^2 - 16}$$

Solution: We know that for $y(t) = \tan t$ we have $y(0) = 0$ and $y'(0) = 1$. By the Existence/Uniqueness Theorem applied to this ODE/IVP, we know that there must exist a unique solution to this IVP with interval at least $(-4, 4)$, but $y(t) = \tan t$ has an asymptote at $t = \frac{\pi}{2}$, so $\tan t$ is cannot be the unique solution guaranteed by the Theorem, since the actual solution must exist on all of $(-4, 4)$.   □

Exercises

In 1-4, find the largest interval (a, b) for which the Existence/Uniqueness Theorem guarantees as a unique solution to the ODE/IVP.

1. (a) $ty^{(3)} + \csc ty' + y = t^2$, $y(1) = 1$, $y'(1) = 2$, $y''(1) = 2$

   (b) $ty^{(3)} + \csc ty' + y = t^2$, $y(10) = 1$, $y'(10) = 2$, $y''(10) = 2$

2. (a) $ty^{(3)} + \dfrac{1}{t-2}y' + \dfrac{1}{t-3}y = t$, $y(1) = 2$, $y'(1) = 2$, $y''(1) = 2$

   (b) $ty^{(3)} + \dfrac{1}{t-2}y' + \dfrac{1}{t-3}y = t$, $y(11) = 2$, $y'(11) = 2$, $y''(11) = 2$

3. $t^3 y^{(3)} + t^4 y' = t^2$, $y(-1) = 1$, $y'(-1) = 2$, $y''(-1) = 2$

4. $\sqrt{2 - t^2}y'' - \sqrt{t}y' = t$, $y(1) = 1$, $y'(1) = 2$, $y''(1) = 2$

4.3 Reduction of Order

Suppose that we were given one particular non-zero solution $y_1(t)$ to a homogenous linear ODE. Then we can reduce the order of the differential equation by one by using a simple trick, namely plug $y(t) = v(t)y_1(t)$ where $v(t)$ is some unknown function. After plugging into the differential equation, all terms that involve $v(t)$ will drop out (only terms that involve its derivatives will survive). The resulting differential equation is of order at most $n - 1$ in the variable $w(t) = v'(t)$. The reason for this is that (by the product rule) any term that only involves $v(t)$ will have coefficient that solves the homogeneous ODE in terms of $y_1(t)$. We demonstrate this with several examples.

Example 4.9 *Verify that $y_1(t) = t^2$ solves the second order linear homogeneous ODE, then find a second solution.*

$$y'' - \frac{2}{t^2}y = 0$$

Solution: We know that for $y_1(t) = t^2$ we have $y_1''(t) = 2$, so plugging in, we see

$$y_1''(t) - \frac{2}{t^2}y_1(t) = 2 - \frac{2}{t^2}(t^2) = 0.$$

So, $y_1(t) = t^2$ solves the DE.

Next, we plug $y(t) = v(t)t^2$ into the left-hand side of the ODE, we obtain:

$$y''(t) - \frac{2}{t^2}y(t) = 2 - \frac{2}{t^2}(t^2)$$

$$= v(t)y_1''(t) + 2v'(t)y_1'(t) + v''(t)y_1(t) - \frac{2}{t^2}(v(t)y_1(t))$$

$$= v(t)\left(y_1''(t) - \frac{2}{t^2}y_1(t)\right) + 2v'(t)y_1'(t) + v''(t)y_1(t)$$

The term in the parentheses is zero (since $y_1(t)$ solves the homogeneous DE), so if we could solve

$$2v'(t)y_1'(t) + v''(t)y_1(t) = 0$$

for $v(t)$ we would be in business. Therefore if we simply label $w = v'(t)$ and substitute $y_1(t) = t^2$, we would have a first order DE namely:

$$4tw(t) + w'(t)t^2 = 0$$

or

$$\frac{dw}{dt} = -\frac{4}{t}w(t).$$

We can easily solve this separable DE, to obtain

$$\ln w(t) = -4\ln t + C$$

or

$$w(t) = Ct^{-4}$$

Therefore, by integrating,

$$v(t) = C_1 t^{-3} + C_2.$$

But then, $y(t) = v(t)y_1(t) = (C_1 t^{-3} + C_2)t^2$ is also a solution. Therefore the general solution to the DE is

$$y(t) = C_1\frac{1}{t} + C_2 t^2.$$

$\square$

There is a general formula for this method if a solution has been found to a second order linear ODE:

2nd Order Reduction of Order

Consider

$$y''(t) + a_1(t)y'(t) + a_0(t)y(t) = 0,$$

and suppose that $y_1(t)$ is a non-trivial solution to this ODE. Then,

$$y_2(t) = y_1(t)\int \frac{1}{[y_1(t)]^2 e^{\int a_1(t)\,dt}}\,dt$$

is a second linearly independent solution.

(Note the $y_1(t)$ in this formula cannot be distributed inside until both integrals are resolved).

Proof: Suppose that $y_1(t)$ is a non-trivial solution to

$$y''(t) + a_1(t)y'(t) + a_0(t)y(t) = 0.$$

Set $y(t) = v(t)y_1(t)$. Plugging into the left side of the ODE and using the product rule we obtain:

$$v(t)y_1''(t) + 2v'(t)y_1'(t) + v''(t)y_1(t) + a_1(t)\left(v(t)y_1'(t) + v'(t)y_1(t)\right) + a_0(t)v(t)y_1(t)$$

$$= v(t)\left(y_1''(t) + a_1(t)y_1'(t) + a_0(t)y_1(t)\right) + 2v'(t)y_1'(t) + v''(t)y_1(t) + a_1(t)v'(t)y_1(t)$$

$$= 2v'(t)y_1'(t) + v''(t)y_1(t) + a_1(t)v'(t)y_1(t)$$

If we set $w(t) = v'(t)$ we get

$$= \frac{dw}{dt}y_1(t) + (a_1(t)y_1(t) + 2y_1'(t))\,w(t)$$

To set this equal to zero amounts to solving a first order linear:

$$\frac{dw}{dt} + \left(a_1(t) + 2\frac{y_1'(t)}{y_1(t)}\right)w(t) = 0$$

Which has solution

$$w(t) = \frac{C}{e^{\int a_1(t) + 2\frac{y_1'(t)}{y_1(t)}\,dt}}$$

Taking $C = 1$ and simplifying,

$$w(t) = \frac{1}{[y_1(t)]^2 e^{\int a_1(t)\,dt}}$$

So

$$v(t) = \int \frac{1}{[y_1(t)]^2 e^{\int a_1(t)\,dt}}\,dt$$

Therefore,

$$y_2(t) = y_1(t)\int \frac{1}{[y_1(t)]^2 e^{\int a_1(t)\,dt}}\,dt$$

Next consider the Wronskian:

$$W\left[y_1(t), y_1(t) \int \frac{1}{[y_1(t)]^2 e^{\int a_1(t)\, dt}}\, dt\right]$$

$$= \begin{vmatrix} y_1(t) & y_1(t) \int \frac{1}{[y_1(t)]^2 e^{\int a_1(t)\, dt}}\, dt \\ y_1'(t) & y_1'(t) \int \frac{1}{[y_1(t)]^2 e^{\int a_1(t)\, dt}}\, dt + \frac{1}{y_1(t) e^{\int a_1(t)\, dt}} \end{vmatrix}$$

which simplifies to $\frac{1}{e^{\int a_1(t)\, dt}}$ which is never zero, so the functions are linearly independent. $\square$

Exercises

In 1-4, find a second linearly independent solution using the given solution to the homogeneous ODE

1. $t^2 y'' - ty' - 3y = 0, \quad y_1(t) = t^3$

2. $t^2 y'' - 6y = 0, \quad y_1(t) = t^3$

3. $y'' + \dfrac{2}{t} y' = 0, \quad y_1(t) = 1$

4. $8t^2 y'' + 2ty' + y = 0, \quad y_1(t) = \sqrt{t}$

4.4 nth Order Linear ODE with Constant Coefficients

In this section we extend the results of second order linear ODE with constant coefficients. We start with the homogeneous case.

---

nth Order Linear Homogeneous with Constant Coefficients

Consider

$$a_n y^{(n)} + a_{n-1} y^{(n-1)} + \ldots + a_2 y'' + a_1 y' + a_0 y = 0,$$

and the associated characteristic polynomial

$$a_n r^n + a_{n-1} r^{n-1} + \ldots + a_2 r^2 + a_1 r + a_0 = 0.$$

- For each real root R of the characteristic polynomial: e^{Rt} is a solution to the the ODE. (If R is repeated k, times, then $e^{Rt}, te^{Rt}, \ldots, t^{k-1}e^{Rt}$ are all solutions to the DE.)

- For each pair of complex roots $\alpha + \beta i$ of the characteristic polynomial: $e^{\alpha t}\cos\beta t, e^{\alpha t}\sin\beta t$ are solutions to the the ODE. (If a pair of complex roots is repeated k, times, then $e^{\alpha t}\cos\beta t, e^{\alpha t}\sin\beta t, te^{\alpha t}\cos\beta t, te^{\alpha t}\sin\beta t, \ldots, t^{k-1}e^{\alpha t}\cos\beta t, t^{k-1}e^{\alpha t}\sin\beta t$ are all solutions to the DE.)

We use the above result together with the superposition principle to obtain general solutions for nth order homogeneous linear ODE with constant coefficients.

Example 4.10 *Solve the ODE*

$$y^{(3)} - 8y = 0$$

Solution: The characteristic polynomial is $r^3 - 8 = 0$, which has root $r = 2$. We use polynomial long division to factor the polynomial in order to find the remaining roots ($r - 2$ must go evenly into $r^3 - 8$ since $r = 2$ is a root).

$$
\begin{array}{r}
r^2 + 2r + 4 \\
r - 2 \overline{\smash{)}\; r^3 \qquad\qquad\qquad - 8} \\
\underline{r^3 - 2r^2} \\
2r^2 \\
\underline{2r^2 - 4r} \\
4r - 8 \\
\underline{4r - 8} \\
0
\end{array}
$$

Thus, we see that $r^3 - 8 = (r-2)(r^2+2r+4)$ and that $r^2+2r+2 = (r+1)^2+3$ has roots $-1 \pm \sqrt{3}i$.

So the general solution of the ODE is

$$
y(t) = c_1 e^{-2t} + c_2 e^{-t} \cos(\sqrt{3}t) + c_3 e^{-t} \sin(\sqrt{3}t)
$$

$\square$

Example 4.11 *Solve the ODE*

$$
y^{(4)} + 2y'' + y = 0
$$

Solution: The characteristic polynomial is $r^4 + 2r^2 + 1 = 0$. Note that $r^4 + 2r^2 + 1 = (r^2 + 1)^2$ But then, $r^2 + 1$ has roots $r = \pm i$, repeated twice.

So the general solution of the ODE is

$$
y(t) = c_1 \cos t + c_2 \sin t + c_3 t \cos t + c_4 t \sin t
$$

$\square$

Exercises

In 1-5, find the general solution to the homogeneous ODE

1. $y^{(3)} + 2y'' + y' = 0$

2. $y^{(3)} + 2y'' - 5y' - 6y = 0$

3. $y^{(3)} + 6y'' + 11y' + 6y = 0$

4. $y^{(3)} + 2y'' + 4y' + 8y = 0$

5. $y^{(4)} + 5y'' + 4y = 0$ (Hint: $r^2 + 1$ is a factor of the characteristic polynomial.)

In 6-7, find the particular solution to the homogeneous ODE

6. $y^{(3)} - y'' + 4y' - 4y = 0$, $y(0) = 1$, $y'(0) = 0$, $y''(0) = 0$

7. $y^{(3)} - 2y'' + 4y' - 8y = 0$, $y(0) = 1$, $y'(0) = 2$, $y''(0) = 0$

4.5   Undetermined Coefficients Revisited

In this section, we address how to solve higher order non-homogeneous linear differential equations. The method is practically identical to the second order case.

Method of Undetermined Coefficients

For a linear ODE with constant coefficients

$$a_n y^{(n)} + a_{n-1} y^{(n-1)} + \ldots + a_2 y'' + a_1 y' + a_0 y = f(t),$$

by the superposition principle, if $y_P(t)$ solves the non-homogeneous and $y_{homog}(t)$ is a general solution for the homogeneous

$$a_n y^{(n)} + a_{n-1} y^{(n-1)} + \ldots + a_2 y'' + a_1 y' + a_0 y = 0,$$

then $y_P(t) + y_{homog}(t)$ is a general solution for the non-homogeneous.

As in the second order case, we guess at the form of one particular solution and plug back into the DE to find the coefficients of the guess.

Below is a chart to assist in determining the appropriate guess:

Undetermined Coefficients Method

The nonhomogeneous second order linear DE with constant coefficients

$$a_n y^{(n)} + a_{n-1} y^{(n-1)} + \ldots + a_2 y'' + a_1 y' + a_0 y = f(t) \qquad (4.5)$$

has a solution of the form:

f(t)	Guess
constant	A
e^{rt}	Ae^{rt}
$\cos(kt)$	$A\cos(kt) + B\sin(kt)$
$\sin(kt)$	$A\cos(kt) + B\sin(kt)$
$c_n t^n + c_{n-1}t^{n-1} + \ldots + c_1 t + c_0$	$A_n t^n + A_{n-1}t^{n-1} + \ldots + A_1 t + A_0$
$(n^{th}\ degree\ poly) \cdot \cos(kt)$ $+(n^{th}\ degree\ poly) \cdot \sin(kt)$	$(A_n t^n + \ldots + A_1 t + A_0)\cos(kt)$ $+(B_n t^n + \ldots + B_1 t + B_0)\sin(kt)$
$(n^{th}\ degree\ poly) \cdot e^{rt}$	$(A_n t^n + \ldots + A_1 t + A_0)e^{rt}$
$(n^{th}\ degree\ poly) \cdot e^{rt} \cdot \cos(kt)$ $+(n^{th}\ degree\ poly) \cdot e^{rt} \cdot \sin(kt)$	$(A_n t^n + \ldots + A_1 t + A_0)e^{rt}\cos(kt)$ $+(B_n t^n + \ldots + B_1 t + B_0)e^{rt}\sin(kt)$

The unknown constants in the guess are obtained by plugging the guess into the DE and solving for the coefficients.

The exception is when terms from the above guess coincide with the homogeneous solution to the DE. In these cases, the guess term needs to be multiplied by the smallest power of t so that the guess no longer has terms that coincide with the general solution of the homogeneous.

Example 4.12 *Find the general solution to*

$$y''' - y'' + 4y' - 4y = 17e^{-3t}$$

Solution: The characteristic polynomial $r^3 - r^2 + 4r - 4 = 0$ factors as $(r-1)(r^2+4)$, so it has roots 1 and $\pm 2i$. Therefore,

$$y_{homog}(t) = c_1 e^t + c_2 \cos(2t) + c_3 \sin(2t).$$

Next we guess at a particular solution to the nonhomogeneous, namely $y_P = Ae^{-3t}$. Note that $(y_P)' = -3Ae^{-3t}$, $(y_P)'' = 9Ae^{-3t}$, and $(y_P)^{(3)} = -27Ae^{-3t}$.

Plugging into the DE we see that

$$-27Ae^{-3t} - 9Ae^{-3t} + 4(-3Ae^{-3t}) - 4(Ae^{-3t}) = 0$$

or

$$-52Ae^{-3t} = 17e^{-3t}$$

therefore $A = -\frac{17}{52}$ and so, the general solution is

$$y_{general}(t) = c_1 e^t + c_2 \cos(2t) + c_3 \sin(2t) - \frac{17}{52}e^{-3t}.$$

$\square$

Exercises

Find ONE solution for each of the DEs

1. $y^{(3)} + y' + 2y = 3e^t$

2. $y^{(3)} + y' + 5y = -t^2$

3. $y^{(4)} - 7y = \cos(4t)$

4. $z^{(3)} - z = e^t$

5. $z^{(4)} - 16z = 6e^{2t}$

6. $z^{(101)} + z' + z = 4e^t$

7. $z^{(101)} + z' + z = 4\cos t$

Find general solutions for each of the DEs

8. $y''' + y' = \sin x$

9. $y^{(3)} - 3y'' + 3y' - y = e^{2t}$

10. $y^{(3)} - 3y'' + 3y' - y = e^{-t}$

11. $y^{(4)} + 6y' - 5y = t^2 + 1$

12. $z'' + 4z' + z = t \sin t$

13. $z^{(4)} - z = t$

14. $z^{(3)} + z'' = 2t - 3$

15. $13z^{(3)} + 7z'' + 6z = e^{-t}$

Find the particular solution to the initial value problems

16. $y^{(3)} + 4y'' = 2e^t, \quad y(0) = 0, \quad y'(0) = 0, \quad y''(0) = 1$

17. $y^{(3)} + y'' + y' + y = t^3, \quad y(0) = 0, \quad y'(0) = 0, \quad y''(0) = 1$

18. $y^{(3)} + y'' + y' + y = e^{-t}, \quad y(0) = 0, \quad y'(0) = 0, \quad y''(0) = 1$

19. $z^{(4)} + 4z'' + 4z = \cos(2t), \quad z(0) = 0, \quad z'(0) = 0, \quad z''(0) = 0, \quad z^{(3)}(0) = 1$

5 Chapter

Laplace Transforms

5.1   Introduction and Definition

In this section we introduce the notion of the Laplace transform. We will use this idea to solve differential equations, but the method also can be used to sum series or compute integrals. We begin with the definition:

Laplace Transform

Let $f(t)$ be a function whose domain includes $(0, \infty)$ then the Laplace transform of $f(t)$ is:

$$\mathcal{L}\left[f(t)\right](s) = \int_0^\infty f(t)e^{-st}\ dt$$

Note that the Laplace transform of a function is another function whose variable is usually denoted as s. Also note that the Laplace transform involves an improper integral. We compute the Laplace transform of several functions:

Example 5.1 *Compute the Laplace transform of $f(t) = 1$*

Solution:

$$\mathcal{L}\left[1\right](s) = \int_0^\infty 1e^{-st}\ dt = \lim_{c \to \infty} \left. -\frac{1}{s}e^{-st}\right|_0^c$$

157

In order for this limit to exist, we must insist that $s \neq 0$ and that $s > 0$ so that e^{-sc} has a limit (of zero). When $s > 0$, we obtain

$$-\frac{1}{s} \lim_{c \to \infty} (e^{-sc} - 1) = \frac{1}{s}$$

So

$$\mathcal{L}\left[1\right](s) = \frac{1}{s}; \quad s > 0.$$

$\square$

Example 5.2 *Compute the Laplace transform of $f(t) = t$*

Solution:

$$\mathcal{L}\left[t\right](s) = \int_0^\infty t e^{-st} \, dt$$

We integrate by Parts (letting $u = t$ and $dv = e^{-st} \, dt$) to obtain:

$$\int t e^{-st} \, dt = -\frac{1}{s} t e^{-st} - \frac{1}{s^2} e^{-st},$$

so

$$\int_0^\infty t e^{-st} \, dt = \lim_{c \to \infty} \left(-\frac{1}{s} t e^{-st} - \frac{1}{s^2} e^{-st} \right) \Bigg|_0^c$$

In order for this limit to exist, we again must insist that $s \neq 0$ and that $s > 0$ so that e^{-sc} has a limit (of zero). We obtain

$$-\frac{1}{s} \lim_{c \to \infty} (c e^{-sc} - 0) - \frac{1}{s^2} \lim_{c \to \infty} (e^{-sc} - 1)$$

which exists for $s > 0$ and after L'Hôpital's rule yields

$$\mathcal{L}\left[t\right](s) = \frac{1}{s^2}; \quad s > 0.$$

$\square$

The previous example can be upgraded to find the Laplace transform of $f(t) = t^n$ for any positive integer n.

Example 5.3 *Show that if $f(t) = t^n$, for any positive integer n then*

$$\mathcal{L}\left[t^n\right](s) = \frac{n}{s}\mathcal{L}\left[t^{n-1}\right](s)$$

Solution:

$$\mathcal{L}\left[t^n\right](s) = \int_0^\infty t^n e^{-st}\,dt$$

We integrate by Parts (letting $u = t^n$ and $dv = e^{-st}\,dt$) to obtain:

$$\int t^n e^{-st}\,dt = -\frac{1}{s}t^n e^{-st} + \frac{1}{s}\int nt^{n-1}e^{-st}\,dt,$$

so

$$\int_0^\infty t^n e^{-st}\,dt = \lim_{c\to\infty}\left(-\frac{1}{s}t^n e^{-st}\right)\Bigg|_0^c + \frac{n}{s}\mathcal{L}\left[t^{n-1}\right](s).$$

In order for this limit to exist, we again must insist that $s \neq 0$ and that $s > 0$ so that e^{-sc} has a limit (of zero). Using L'Hôpital's rule n times, we see that

$$\lim_{c\to\infty}\left(-\frac{1}{s}c^n e^{-sc}\right) = 0.$$

$\square$

Using this result inductively, we compute:

$$\mathcal{L}\left[t^2\right](s) = \frac{2}{s}\mathcal{L}[t](s) = \frac{2}{s}\cdot\frac{1}{s^2} = \frac{2}{s^3}$$

$$\mathcal{L}\left[t^3\right](s) = \frac{3}{s}\mathcal{L}\left[t^2\right](s) = \frac{3}{s}\cdot\frac{2}{s}\cdot\frac{1}{s^2} = \frac{6}{s^4}$$

$$\mathcal{L}\left[t^4\right](s) = \frac{4}{s}\mathcal{L}\left[t^3\right](s) = \frac{4}{s}\cdot\frac{3}{s}\cdot\frac{2}{s}\cdot\frac{1}{s^2} = \frac{24}{s^5}$$

Note that in all cases above, we must have $s > 0$. In summary,

---

Laplace Transform of single term monic polynomials

$$\mathcal{L}\left[t^n\right](s) = \frac{n!}{s^{n+1}}; \quad s > 0$$

Linear Combinations and Laplace Transform

Theorem 5.4 *Let $f(t)$ and $g(t)$ be functions with Laplace transforms $F(s)$ and $G(s)$ respectively then for any constants a and b*

$$\mathcal{L}\left[af(t) + bg(t)\right](s) = aF(s) + bG(s)$$

Proof:

$$\mathcal{L}\left[af(t) + bg(t)\right](s) = \int_0^\infty [af(t) + bg(t)]e^{-st}\ dt$$

$$= \int_0^\infty af(t)e^{-st}\ dt + \int_0^\infty bg(t)e^{-st}\ dt = aF(s) + bG(s)$$

$\square$

From this result we can take the Laplace transform of any arbitrary polynomial in the next example,

Example 5.5 *Find $\mathcal{L}\left[4t^3 + 8t^2 - 7\right](s)$*

Solution:

$$\mathcal{L}\left[4t^3 + 8t^2 - 7\right](s) = 4\mathcal{L}\left[t^3\right](s) + 8\mathcal{L}\left[t^2\right](s) - 7\mathcal{L}\left[1\right](s)$$

$$= 4\left(\frac{6}{s^4}\right) + 8\left(\frac{2}{s^3}\right) - 7\frac{1}{s}$$

$$= \frac{24}{s^4} + \frac{16}{s^3} - \frac{7}{s},$$

where $s > 0$. $\square$

Laplace Transform of exponential functions

$$\mathcal{L}\left[e^{at}\right](s) = \frac{1}{s - a}; \quad s > a$$

Proof:

$$\mathcal{L}\left[e^{at}\right](s) = \int_0^\infty e^{at}e^{-st}\,dt$$

$$= \int_0^\infty e^{(a-s)t}\,dt,$$

which can be integrated with respect to t (use $u = (a-s)t$ and $du = (a-s)\,dt$). We obtain

$$= \lim_{b\to\infty} \frac{1}{a-s}e^{(a-s)b} - \frac{1}{a-s}$$

The above limit exists exactly when $s > a$, so

$$\mathcal{L}\left[e^{at}\right](s) = \frac{1}{s-a}, \quad s > a$$

$\square$

Example 5.6 *Find*

$$\mathcal{L}\left[2^t\right](s)$$

Solution: Rewriting $2^t = e^{ln2^t} = e^{tln2}$ and taking $a = \ln 2$,

$$\mathcal{L}\left[e^{(ln2)t}\right](s) = \frac{1}{s - \ln 2}, \quad s > \ln 2$$

$\square$

Lastly,

Laplace Transform of sine and cosine
$\mathcal{L}\left[\cos(\beta t)\right](s) = \dfrac{s}{s^2 + \beta^2}, \quad s > 0$
$\mathcal{L}\left[\sin(\beta t)\right](s) = \dfrac{\beta}{s^2 + \beta^2}, \quad s > 0$

We derive the second formula and leave the derivation of the first formula as an exercise. By definition:

$$\mathcal{L}\left[\sin(\beta t)\right](s) = \int_0^\infty e^{-st}\sin(\beta t)\,dt.$$

Using integration by parts with $u = \sin(\beta t)$ and $dv = e^{-st}\,dt$ we obtain

$$\int_0^\infty e^{-st}\sin(\beta t)\,dt = -\frac{1}{s}\sin(\beta t)e^{-st}\Big|_0^\infty - \int_0^\infty(-\frac{\beta}{s})e^{-st}\cos(\beta t)\,dt.$$

Using parts again with $u = \cos(\beta t)$ and $dv = e^{-st}\,dt$, we obtain

$$= -\frac{1}{s}\sin(\beta t)e^{-st}\Big|_0^\infty + (\frac{\beta}{s})\left[(\frac{-1}{s})e^{-st}\cos(\beta t)\Big|_0^\infty - \int_0^\infty(\frac{\beta}{s})e^{-st}\sin(\beta t)\,dt\right]$$

From the squeeze theorem we see that for $s > 0$ $\lim\limits_{c\to\infty} e^{-sc}\cos(\beta c) = 0$ and $\lim\limits_{c\to\infty} e^{-sc}\sin(\beta c) = 0$, so the above reduces to

$$= (\frac{\beta}{s^2}) - (\frac{\beta^2}{s^2})\int_0^\infty(e^{-st}\sin(\beta t)\,dt$$

Now recall that the left-hand side of this equation (what we were solving for) is

$$\int_0^\infty e^{-st}\sin(\beta t)\,dt$$

so we have

$$\int_0^\infty e^{-st}\sin(\beta t)\,dt = \left(\frac{\beta}{s^2}\right) - \left(\frac{\beta^2}{s^2}\right)\int_0^\infty e^{-st}\sin(\beta t)\,dt$$

So moving the term with the integral on the right side to the left side gives

$$\int_0^\infty e^{-st}\sin(\beta t)\,dt + \left(\frac{\beta^2}{s^2}\right)\int_0^\infty e^{-st}\sin(\beta t)\,dt = (\frac{\beta}{s^2})$$

Factoring,

$$\left(1 + \frac{\beta^2}{s^2}\right)\int_0^\infty e^{-st}\sin(\beta t)\,dt = (\frac{\beta}{s^2})$$

So

$$\int_0^\infty e^{-st}\sin(\beta t)\,dt = \frac{(\frac{\beta}{s^2})}{\left(1 + \frac{\beta^2}{s^2}\right)} = \frac{\beta}{s^2 + \beta^2}$$

NOTE: It is customary to use the independent variable s for a function that is an output of a Laplace transform and the independent variable t for a function that is an output of a Laplace transform. This convention will be handy in the later sections.

The below result gives a condition that guarantees the existence of the Laplace transform.

A condition that guarantees the existence of $\mathcal{L}\left[f(t)\right](s)$

Suppose that there are numbers α and M so that

$$|f(t)| \le Me^{\alpha t}$$

for all $t > 0$. Then the Laplace transform of $f(t)$ exists with a domain of at least $s > \alpha$.

Proof: Suppose that there are numbers α and M so that

$$|f(t)| \le Me^{\alpha t}$$

for all $t > 0$. Then

$$\mathcal{L}(|f(t)|)(s) = \lim_{b \to \infty} \int_0^b e^{-st}|f(t)|\ dt$$

$$\le \int_0^\infty e^{-st}Me^{\alpha t}\ dt = M\mathcal{L}[e^{\alpha t}](s) = \frac{M}{s - \alpha},$$

for $s > \alpha$.

Therefore, since

$$\int_0^b e^{-st}|f(t)|\ dt$$

is increasing in b and bounded for all b,

$$\lim_{b \to \infty} \int_0^b e^{-st}|f(t)|\ dt$$

exists. This implies that

$$\lim_{b \to \infty} \int_0^b e^{-st}f(t)\ dt$$

also must exist. □

Note that most all exponential functions, polynomials, and the trig functions sine and cosine satisfy this condition but $\ln x$, $\tan x$ and e^{t^2} do not. These functions do not have Laplace transforms.

Exercises

1. Compute $\mathcal{L}[f(t)](s)$ for $f(t) = 0$.

2. Compute $\mathcal{L}[f(t)](s)$ for

$$f(t) = \begin{cases} 1 & \text{if } t < 8 \\ 0 & \text{if } t \geq 8. \end{cases}$$

3. Compute $\mathcal{L}[f(t)](s)$ for

$$f(t) = \begin{cases} t & \text{if } t < 1 \\ 2 - t & \text{if } 1 \leq t < 2 \\ 0 & \text{if } t \geq 2. \end{cases}$$

4. Compute the formula for $\mathcal{L}[\cosh(\beta t)](s)$, where β is a constant.

5. Compute the formula for $\mathcal{L}[\cos(\beta t)](s)$, where β is a constant.

6. Compute $\mathcal{L}[\sinh(\beta t)](s)$,

7. Find the Laplace transform of

$$f(t) = \begin{cases} t - 8 & \text{if } t < 8 \\ e^{t+6} & \text{if } 8 < t < 10 \\ t? & \text{if } 10 < t < 11 \\ 0 & \text{if } t > 11. \end{cases}$$

In 8-13, take the Laplace transform of the function $f(t)$ using the results in this section (do not derive using the definition)

8. $f(t) = t^3 - 7t^2 + 8$

9. $f(t) = 4\sin t - 3\cos t$

10. $f(t) = \cos(2t) - e^{9t}$

11. $f(t) = 1 + 7\sin(5t)$

12. $f(t) = \cosh(\beta t)$, where β is a constant. (Recall $\cosh(z) = \frac{e^z + e^{-z}}{2}$.)

13. $f(t) = \sinh(\beta t)$, where β is a constant. (Recall $\sinh(z) = \frac{e^z - e^{-z}}{2}$).

In 14-16, use the definition to take the Laplace transform of the function $f(t)$ using integration by parts and formulas given in this section

14. $f(t) = te^t$

15. $f(t) = t\sin(2t)$

16. $f(t) = t^2\cos(t)$

17. Recall that $\mathcal{L}[f(t) + g(t)] = \mathcal{L}[f(t)] + \mathcal{L}[g(t)]$. Is it true in general that $\mathcal{L}[f(t) \cdot g(t)] = \mathcal{L}[f(t)] \cdot \mathcal{L}[g(t)]$?

    (a) Try $f(t) = t$ and $g(t) = 1$ to see if $\mathcal{L}[f(t) \cdot g(t)] = \mathcal{L}[f(t)] \cdot \mathcal{L}[g(t)]$.

    (b) Is it ever true that $\mathcal{L}[f(t) \cdot g(t)] = \mathcal{L}[f(t)] \cdot \mathcal{L}[g(t)]$?

5.2 The Inverse Laplace Transform and ODEs

In this section we will see how the Laplace transform can be used to solve differential equations. The key result that allows us to do this is the following:

Laplace Transform of $y'(t)$

Suppose that $\mathcal{L}[y(t)](s)$ exists and that $y(t)$ is differentiable $(0, \infty)$ with derivative $y'(t)$ then

$$\mathcal{L}[y'(t)](s) = s\mathcal{L}[y(t)](s) - y(0)$$

Proof: By definition,

$$\mathcal{L}[y'(t)](s) = \int_0^\infty e^{-st} y'(t) \, dt$$

Setting $u = e^{-st}$ and $dv = y'(t) \, dt$ and using integration by parts, we obtain

$$\mathcal{L}[y'(t)](s) = e^{-st} y(t) \Big|_0^\infty - \int_0^\infty (-s) e^{-st} y(t) \, dt$$

For s big enough, $\lim_{c \to \infty} e^{-sc} y(c) = 0$, since the Laplace transform of $y(t)$ exists (this follows since the continuous integrand of a convergent improper integral must tend to zero – we will not prove this fact here.)

So we obtain

$$\mathcal{L}[y'(t)](s) = -y(0) + s\mathcal{L}[y(t)](s)$$

$\square$

From this result, we derive:

Laplace Transform of $y''(t)$

Suppose that $\mathcal{L}[y(t)](s)$ exists and that $y(t)$ is twice differentiable on $(0, \infty)$ with second derivative $y''(t)$ then

$$\mathcal{L}[y''(t)](s) = s^2 \mathcal{L}[y(t)](s) - sy(0) - y'(0)$$

Proof: We simply use the previous result twice

$$\mathcal{L}[y''(t)](s) = s\mathcal{L}[y'(t)](s) - y'(0) = s\left[s\mathcal{L}[y(t)](s) - y(0)\right] - y'(0).$$

$\square$

Example 5.7 *Solve $y''(t) = t$ using the Laplace Transform*

Solution: Taking the Laplace transform of both sides we obtain

$$\mathcal{L}[y''(t)](s) = \mathcal{L}[t]$$

$$\mathcal{L}[y''(t)](s) = \frac{1}{s^2}$$

so

$$s^2\mathcal{L}[y(t)](s) - sy(0) - y'(0) = \frac{1}{s^2}$$

Solving for $\mathcal{L}[y(t)](s)$ we obtain

$$\mathcal{L}(y(t))(s) = \frac{1}{s^4} + \frac{y(0)}{s} + \frac{y'(0)}{s^2}$$

Now if we could reverse engineer which function $y(t)$ has Laplace transform equal to the right hand side, then we would be done.

By playing a bit, we can see that $y(t) = \frac{1}{6}t^3 + \frac{y(0)}{2} + y'(0)t$ has this Laplace transform. So voila! We have solved the differential equation.

(Note that the general solution obtained from undetermined coefficients would be $y(t) = \frac{1}{6}t^3 + C_1 + C_2t$ which agrees with our solution since $y(0)$ and $y'(0)$ are unspecified constants.) □

The above example illustrates a common checklist for solving DEs using the Laplace transform:

1. Transform the original problem to one involving $\mathcal{L}(y(t))(s)$,

2. Solve for $\mathcal{L}[y(t)](s)$,

3. Undo the Laplace transform to recover $y(t)$.

Definition 5.8 *A continuous function $f(t)$ is the **Inverse Laplace Transform** of a function $F(s)$ if $\mathcal{L}[f(t)](s) = F(s)$. In this case, we write $f(t) = \mathcal{L}^{-1}[F(s)]$*

It is customary shorthand notation to denote $F(s)$ to be the Laplace transform of $f(t)$ and $G(s)$ to be the Laplace transform of $g(t)$. For constants a, b, since

$$\mathcal{L}[af(t) + bg(t)] = a\mathcal{L}[f(t)] + b\mathcal{L}[g(t)] = aF(s) + bG(s)$$

we see that

$$\mathcal{L}^{-1}[aF(s) + bG(s)] = af(t) + bg(t)$$

or

$$\mathcal{L}^{-1}[aF(s) + bG(s)] = a\mathcal{L}^{-1}[F(s)] + b\mathcal{L}^{-1}[G(s)],$$

so the Laplace transform of a linear combination is the linear combination of the Laplace transforms.

The following table lists several inverses which come from the previous section:

<table>
<tr><td colspan="2" align="center">Some Basic Inverse Laplace Transform Facts</td></tr>
<tr><td>$F(s)$</td><td>$\mathcal{L}^{-1}[F(s)]$</td></tr>
<tr><td>$\frac{1}{s}$</td><td>1</td></tr>
<tr><td>$\frac{n!}{s^{n+1}}$</td><td>t^n</td></tr>
<tr><td>$\frac{1}{s-a}$</td><td>e^{at}</td></tr>
<tr><td>$\frac{s}{s^2+\beta^2}$</td><td>$\cos \beta t$</td></tr>
<tr><td>$\frac{\beta}{s^2+\beta^2}$</td><td>$\sin \beta t$</td></tr>
<tr><td>$aF(s) + bG(s)$</td><td>$af(t) + bg(t)$</td></tr>
</table>

Example 5.9 *Find* $\mathcal{L}^{-1}\left[\frac{7}{s^3}\right]$

Solution:

$$\mathcal{L}^{-1}\left[\frac{7}{s^3}\right] = 7\mathcal{L}^{-1}\left[\frac{1}{s^3}\right]$$

We multiply the inner fraction by $\frac{2}{2}$ to obtain the form in the previous table.

$$= 7\mathcal{L}^{-1}\left[\frac{2}{2s^3}\right] = \frac{7}{2}\mathcal{L}^{-1}\left[\frac{2}{s^3}\right]$$

$$= \frac{7}{2}t^2$$

$\square$

Example 5.10 *Find* $\mathcal{L}^{-1}\left[\frac{2s-1}{s^2+4}\right]$

Solution:

$$\mathcal{L}^{-1}\left[\frac{2s-1}{s^2+4}\right] = 2\mathcal{L}^{-1}\left[\frac{s}{s^2+4}\right] - \mathcal{L}^{-1}\left[\frac{1}{s^2+4}\right]$$

Again, the inner fraction of the second must be multiplied by $\frac{2}{2}$ to put it in the form in the table ($\beta = 2$ since $\beta^2 = 4$).

$$= 2\cos 2t - \mathcal{L}^{-1}\left[\frac{2}{2(s^2+4)}\right]$$

$$= 2\cos 2t - \frac{1}{2}\mathcal{L}^{-1}\left[\frac{2}{s^2+4}\right]$$

$$= 2\cos 2t - \frac{1}{2}\sin 2t$$

$\square$

Example 5.11 *Solve $y''(t) + 9y(t) = e^{4t}$ with $y(0) = 1$ and $y'(0) = 0$*

Solution: We first take the Laplace transform of both sides to obtain

$$\mathcal{L}[y''(t) + 9y(t)] = \mathcal{L}[e^{4t}]$$

$$s^2\mathcal{L}[y(t)] - sy(0) - y'(0) + 9\mathcal{L}[y(t)] = \frac{1}{s-4}$$

$$(s^2+9)\mathcal{L}[y(t)] = \frac{1}{s-4} + s$$

$$\mathcal{L}[y(t)] = \frac{1}{(s-4)(s^2+9)} + \frac{s}{s^2+9}$$

We use partial fractions to decompose the first term into a sum of terms that are recognizable inverses.

Noting that by partial fractions:

$$\frac{1}{(s-4)(s^2+9)} = \frac{A}{s-4} + \frac{Bs+C}{s^2+9},$$

where $A = \frac{1}{25}$ $B = -\frac{1}{25}$ $C = -\frac{4}{25}$. Substituting in, we obtain

$$\mathcal{L}[y(t)] = \frac{1}{25}\left(\frac{1}{s-4}\right) - \frac{1}{25}\left(\frac{s}{s^2+9}\right) - \frac{4}{25}\left(\frac{1}{s^2+9}\right) + \frac{s}{s^2+9}$$

$$\mathcal{L}[y(t)] = \frac{1}{25}\left(\frac{1}{s-4}\right) + \frac{24}{25}\left(\frac{s}{s^2+9}\right) - \frac{4}{25}\left(\frac{1}{s^2+9}\right)$$

So

$$y(t) = \frac{1}{25}e^{4t} + \frac{24}{25}\cos 3t - \frac{4}{75}\sin 3t$$

(note that the '3' appeared in the denominator since $\beta = 3$ and in order to find $\mathcal{L}^{-1}\left[\frac{1}{s^2+9}\right]$ we multiply the inner fraction by $\frac{3}{3}$.) □

We end this section by noting the following extension

Laplace Transform of $y^{(n)}(t)$

Suppose that $\mathcal{L}[y(t)](s)$ exists and that $y(t)$ is differentiable n times on $(0, \infty)$ with n^{th} derivative $y^{(n)}(t)$ then

$$\mathcal{L}[y^{(n)}(t)](s) = s^n\mathcal{L}[y(t)](s) - s^{n-1}y(0) - s^{n-2}y'(0) - \ldots - sy^{(n-2)}(0) - y^{(n-1)}(0)$$

Exercises

In 1-7, find the Inverse Laplace transform of the function $F(s)$. You may need partial fractions

1. $F(s) = \frac{1}{s+2}$

2. $F(s) = \frac{1}{s^4}$

3. $F(s) = \frac{s+5}{s^2+16}$

4. $F(s) = \frac{4}{s+6}$

5. $F(s) = \frac{1}{(s+1)(s^2+1)}$

6. $F(s) = \frac{s}{(s+1)(s^2+1)}$

7. $F(s) = \frac{3}{s^3+s}$

8. Use the Laplace transform to find the general solution of
$$y'' + 6y' + 5y = t$$

9. Use the Laplace transform to find the general solution of
$$y'' - y = e^t$$

10. Use the Laplace transform to find the general solution of
$$y'' + y = e^t$$

11. Use the Laplace transform to find the solution of the IVP
$$y'' + y = 2, y(0) = 1, \quad y'(0) = -1$$

12. Use the Laplace transform to find the solution of the IVP
$$y'' + 3y' + 2y = 2, y(0) = 1, \quad y'(0) = -1$$

In 13-16, solve the ODE/IVP using the Laplace Transform

13. $y'' + 2y' + y = 0, \quad y(0) = 1, \quad y'(0) = 2$

14. $y'' + 4y' + 3y = t^2, \quad y(0) = 1, \quad y'(0) = 0$

15. $y'' + 6y' + 5y = \sin t, \quad y(0) = 0, \quad y'(0) = 0$

16. $y'' - 3y' + 4y = e^t, \quad y(0) = 1, \quad y'(0) = 0$

5.3   Useful Properties of the Laplace Transform

In this section, we look at several theorems which can be used to solve DEs using the Laplace transform method. We start with:

Laplace Transform Exponential Shift Theorem (Forward)

$$\mathcal{L}[e^{at}f(t)](s) = \mathcal{L}[f(t)](s-a) = F(s-a)$$

Effectively, this says $\mathcal{L}[e^{at}f(t)](s)$ is equal to $\mathcal{L}[f(t)](w)$, (using w for the traditional s in the definition) and replacing $w = s - a$.

Proof: By definition:

$$\mathcal{L}[e^{at}f(t)] = \int_0^\infty e^{at}e^{-st}f(t)\ dt$$

$$= \int_0^\infty e^{-(s-a)t}f(t)\ dt.$$

Also by definition,

$$\mathcal{L}[f(t)](s-a) = \int_0^\infty e^{-(s-a)t}f(t)\ dt.$$

$\square$

We can use this result to compute Laplace transforms of exponentials multiplied by functions whose Laplace transforms we already know.

Example 5.12 *Find $\mathcal{L}[e^{-2t}\cos(3t)](s)$*

Solution: By the shifting theorem

$$\mathcal{L}[e^{-2t}\cos(3t)](s) = \mathcal{L}[\cos(3t)](s-(-2)) = \mathcal{L}[\cos(3t)](s+2) =$$

The Laplace transform of $\cos(3t)$ with (different) variable w is

$$\mathcal{L}[\cos(3t)](w) = \frac{w}{w^2 + 3^2},$$

so, replacing w with $s + 2$,

$$= \mathcal{L}[\cos(3t)](s+2) = \frac{s+2}{(s+2)^2 + 9}.$$

$$= \frac{s+2}{s^2 + 4s + 13}.$$

Further note that in order for $\mathcal{L}[\cos(3t)](w)$ to exist, $w > 0$ so the domain is $s + 2 > 0$ or $s > -2$. □

Example 5.13 *Find* $\mathcal{L}[t^3 e^t](s)$

Solution: By the shifting theorem

$$\mathcal{L}[t^3 e^t](s) = \mathcal{L}[t^3](w) = \frac{6}{w^4},$$

where $w = s - 1$, so

$$\mathcal{L}[t^3 e^t](s) = \frac{6}{(s-1)^4}$$

□

Similarly, we obtain the following result:

Laplace Transform Exponential Shift Theorem (Backward)

$$\mathcal{L}^{-1}[F(s - a)]) = e^{at} f(t),$$

where $F(w)$ is the Laplace transform of $f(t)$ (with variable w).

To apply this result, we look for familiar outputs of Laplace transforms with the variable s replaced by a shift.

Example 5.14 *Find* $\mathcal{L}^{-1}\left[\frac{24}{(s-8)^5}\right]$

Solution: This is just a shift of

$$\frac{24}{w^5}$$

which is the Laplace transform of t^4 (with variable $w0$. So $\frac{24}{(s-8)^5} = F(s-8)$ where F is the Laplace transform of $f(t) = t^4$. Therefore, by the Exponential Shifting Theorem

$$\mathcal{L}^{-1}[F(s-8)] = e^{8t}t^4.$$

□

Example 5.15 *Find* $\mathcal{L}^{-1}\left[\frac{s}{s^2+2s+5}\right]$

Solution: The trick here is to complete the square in the denominator and then use the backward shift theorem.

$$\frac{s}{s^2 + 2s + 5} = \frac{s}{s^2 + 2s + 1 + 4} = \frac{s}{(s+1)^2 + 4}$$

We also need an $s+1$ in the numerator to use the backwards shift with cosine, so we write the numerator as:

$$= \frac{s+1-1}{(s+1)^2 + 4} = \frac{s+1}{(s+1)^2 + 4} - \frac{1}{(s+1)^2 + 4}$$

Now we can take the inverse of each term using the backward shift theorem to get

$$\mathcal{L}^{-1}\left[\frac{s}{s^2 + 2s + 5}\right] = \mathcal{L}^{-1}\left[\frac{s+1}{(s+1)^2 + 4}\right] - \mathcal{L}^{-1}\left[\frac{1}{(s+1)^2 + 4}\right]$$

$$= e^{-t}\cos 2t - \frac{1}{2}e^{-t}\sin 2t$$

□

The next result equates differentiation of the Laplace transform F with respect to s to the Laplace transform of the product $tf(t)$.

> ### Differentiation Theorem
> Let $F(s)$ be the Laplace transform of $f(t)$. Then
>
> $$\frac{d}{ds}[F(s)] = -\mathcal{L}[tf(t)](s)$$
>
> or
>
> $$-\frac{d}{ds}[F(s)] = \mathcal{L}[tf(t)](s)$$

Proof: By definition:

$$F'(s) = \lim_{\Delta s \to 0} \frac{F(s + \Delta s) - F(s)}{\Delta s}$$

Note that

$$F(s+\Delta s)-F(s) = \int_0^\infty e^{-(s+\Delta s)t}f(t)\,dt - \int_0^\infty e^{-st}f(t)\,dt = \int_0^\infty [e^{-(s+\Delta s)t}-e^{-st}]f(t)\,dt$$

So

$$\frac{F(s+\Delta s) - F(s)}{\Delta s} = \int_0^\infty \frac{e^{-(s+\Delta s)t} - e^{-st}}{\Delta s}f(t)\,dt$$

Therefore

$$\lim_{\Delta s \to 0} \frac{F(s+\Delta s) - F(s)}{\Delta s} = \int_0^\infty \lim_{\Delta s \to 0} \frac{e^{-(s+\delta s)t} - e^{-st}}{\Delta s}f(t)\,dt$$

Noting that $\lim_{\Delta s \to 0} \frac{e^{-(s+\Delta s)t}-e^{-st}}{\Delta s}$ is the definition of $\frac{d}{ds}(e^{-st})$, we know this limit is $-te^{-st}$.

Substituting, we obtain

$$F'(s) = \int_0^\infty (-te^{-st})f(t)\,dt.$$

By definition, the right hand side is $\mathcal{L}[-tf(t)](s)$ □

We provide several examples which illustrate how to use this result.

Example 5.16 *Find* $\mathcal{L}\left[te^{2t}\right]$

Solution: By the Theorem,

$$\mathcal{L}[tf(t)](s) = -F'(s).$$

Here $f(t) = e^{2t}$, so $F(s) = \frac{1}{s-2}$.
Now

$$F'(s) = -(s-2)^{-2}$$

so

$$\mathcal{L}[te^{2t}](s) = -(-(s-2)^{-2}) = \frac{1}{(s-2)^2}$$

□

Note that we could have also obtained this answer by using the Forward Exponential Shift Theorem.

Example 5.17 *Find* $\mathcal{L}\left[t\cos(3t)\right]$

By the Theorem,

$$\mathcal{L}[tf(t)](s) = -F'(s).$$

Here $f(t) = \cos(3t)$, so $F(s) = \frac{s}{s^2+9}$.
Now

$$F'(s) = \frac{(s^2+9) - s(2s)}{(s^2+9)^2} = \frac{9-s^2}{(s^2+9)^2}$$

so

$$\mathcal{L}[t\cos(3t)](s) = -\frac{9-s^2}{(s^2+9)^2} = \frac{s^2-9}{(s^2+9)^2}$$

□

Next we solve several differential equations using these new results.

Example 5.18 *Find the general solution to* $y'' + 6y' + 6y = \sin 2t$

Solution: Taking the Laplace transform, we obtain

$$\mathcal{L}[y'' + 6y' + 10y] = \mathcal{L}[\sin 2t]$$

so

$$s^2\mathcal{L} - sy(0) - y'(0) + 6s\mathcal{L} - 6y(0) + 10\mathcal{L} = \frac{2}{s^2 + 4}$$

(here $\mathcal{L}$ is shorthand for $\mathcal{L}[y(t)](s)$), so

$$(s^2 + 6s + 10)\mathcal{L} = sy(0) + y'(0) + 6y(0) + \frac{2}{s^2 + 4}$$

Solving for $\mathcal{L}$,

$$\mathcal{L} = \frac{sy(0) + y'(0) + 6y(0) + \frac{2}{s^2+4}}{s^2 + 6s + 10}$$

$$= \frac{sy(0)}{s^2 + 6s + 10} + \frac{y'(0) + 6y(0)}{s^2 + 6s + 10} + \frac{2}{(s^2 + 4)(s^2 + 6s + 10)} \tag{5.1}$$

We now have to use the inverse Laplace transform. We start with the first term, which we will have to write as a shift by completing the square of the denominator. In particular the first term is

$$\frac{sy(0)}{(s + 3)^2 + 1}$$

which we can rewrite as

$$\frac{sy(0)}{(s + 3)^2 + 1} = \frac{(s + 3 - 3)y(0)}{(s + 3)^2 + 1} = \frac{(s + 3)y(0)}{(s + 3)^2 + 1} - \frac{3y(0)}{(s + 3)^2 + 1}.$$

We can invert this using the shift theorem, so the inverse of the first term is:

$$y(0)e^{-3t}\cos t - 3y(0)e^{-3t}\sin t$$

Similarly, the second term in **??** can be realized as a shift, It is

$$\frac{y'(0) + 6y(0)}{(s + 3)^2 + 1}$$

which has inverse transform

$$(y'(0) + 6y(0))e^{-3t}\sin t.$$

Lastly, we use partial fractions on the final term to get

$$\frac{2}{(s^2+4)(s^2+6s+10)} = \frac{As+B}{s^2+4} + \frac{Cs+D}{s^2+6s+10}$$

where $A = -\frac{1}{15}, B = \frac{1}{15}, C = \frac{1}{15}, D = \frac{1}{3}$. (The reader should be familiar with the method of Partial Fractions, normally covered in a second semester calculus class).

We again can easily invert the first term

$$\frac{As+B}{s^2+4} = A\frac{s}{s^2+4} + \frac{B}{2}\frac{2}{s^2+4}$$

which inverts to

$$A\cos 2t + \frac{B}{2}\sin 2t.$$

The second term, after again setting up for the shifting theorem by completing the square, becomes

$$\frac{Cs+D}{s^2+6s+10} = \frac{C(s+3-3)}{(s+3)^2+1} + \frac{D}{(s+3)^2+1}$$

$$= C\frac{(s+3)}{(s+3)^2+1} + (D-3C)\frac{1}{(s+3)^2+1}$$

Taking the inverse, we obtain:

$$= Ce^{-3t}\cos t + (D-3C)e^{-3t}\sin t$$

Putting it altogether,

$$y(t) = y(0)e^{-3t}\cos t - 3y(0)e^{-3t}\sin t + (y'(0)+6y(0))e^{-3t}\sin t$$

$$+Ce^{-3t}\cos t + (D-3C)e^{-3t}\sin t$$

$$-\frac{1}{15}\cos 2t + \frac{1}{30}\sin 2t,$$

where $C = \frac{1}{15}, D = \frac{1}{3}$.
Note that this simplifies to

$$y(t) = (y(0)+C)e^{-3t}\cos t + ((D-3C)+y'(0)+3y(0))e^{-3t}\sin t$$

$$-\frac{1}{15}\cos 2t + \frac{1}{30}\sin 2t.$$

If we had used the method of undetermined coefficients, we would have obtained (possibly in a lot less time)

$$y(t) = C_1 e^{-3t}\cos t + C_2 e^{-3t}\sin t - \frac{1}{15}\cos 2t + \frac{1}{30}\sin 2t.$$

$\square$

The previous example could (and probably should) have been solved using other methods. The next example is one where these other methods would not apply.

Example 5.19 *Find the solution to*

$$y'' + ty' - 2y = 2, \quad y(0) = 0, \quad y'(0) = 0$$

Solution: We take the Laplace transform of both sides of the differential equation.

$$\mathcal{L}[y''] + \mathcal{L}[ty'] - 2\mathcal{L}[y] = \frac{2}{s},$$

$$s^2\mathcal{L}[y] - \frac{d}{ds}(\mathcal{L}[y']) - 2\mathcal{L}[y] = \frac{2}{s},$$

$$s^2\mathcal{L}[y] - \frac{d}{ds}(s\mathcal{L}[y]) - 2\mathcal{L}[y] = \frac{2}{s},$$

Using the product rule, note $\frac{d}{ds}(s\mathcal{L}[y]) = s\mathcal{L}'[y] + \mathcal{L}[y]$ (recall: $\mathcal{L}[y]$ is also a function of s). So,

$$s^2\mathcal{L}[y] - s\mathcal{L}'[y] - \mathcal{L}[y] - 2\mathcal{L}[y] = \frac{2}{s},$$

or

$$(s^2 - 3)\mathcal{L}[y] - s\mathcal{L}'[y] = \frac{2}{s},$$

This is a first order linear differential equation in $\mathcal{L}$ with variable s!

$$\mathcal{L}'[y] + \left(-s + \frac{3}{s}\right)\mathcal{L}[y] = -\frac{2}{s^2},$$

This has integrating factor $e^{-\frac{s^2}{2}+3\ln s}$ which simplifies to $s^3 e^{-\frac{s^2}{2}}$, so

$$\mathcal{L}[y] = \frac{\int s^3 e^{-\frac{s^2}{2}}\left(-\frac{2}{s^2}\right)\,ds + C}{s^3 e^{-\frac{s^2}{2}}}$$

This resolves to

$$\frac{2e^{-\frac{s^2}{2}} + C}{s^3 e^{-\frac{s^2}{2}}}$$

We take $C = 0$ and obtain $\mathcal{L}[y] = \frac{2}{s^3}$ (all other choices do not give valid outputs of Laplace transforms).

Which implies that $y(t) = t^2$ solves the DE. (One may easily check that, indeed $y(t) = t^2$ does solve the DE/IVP.)   □

Exercises

In 1-6, solve the ODE/IVP using the Laplace Transform

1. $y'' + 2y' + 3y = 0$, $y(0) = 1$, $y'(0) = 0$

2. $y'' + 2y' + 3y = t^2$, $y(0) = 1$, $y'(0) = 0$

3. $y'' + 6y' + 10y = \sin t$, $y(0) = 0$, $y'(0) = 0$

4. $y'' + 4y' + 8y = e^t$, $y(0) = 1$, $y'(0) = 0$

5. $y'' + 3ty' - 6y = 6$, $y(0) = 0$, $y'(0) = 0$

6. $y'' + ty' - 3y = -2t$, $y(0) = 0$, $y'(0) = 0$

7. $ty'' - y' + y = t^2$, $y(0) = 0$, $y'(0) = 0$

5.4   Unit Step Functions and Periodic Functions

In this section we will see that we can use the Laplace transform to solve a new class of problems efficiently. In particular, we will be able to consider discontinuous forcing functions. First, we make a definition.

The Unit Step Function
Let $$u(t) = \begin{cases} 0 & t < 0 \\ 1 & t > 0 \end{cases}$$

This function is also called a Heaviside function.

Example 5.20 *Plot the graphs of (a) $u(t)$, (b) $u(t-1)$, (c) $u(t) - u(t-1)$ (d) $(\sin t)\,[u(t) - u(t-1)]$*

Solution: ☐

Note that the general plot of $u(t-a) - u(t-b)$, where $a < b$ is shown in the plot below:

☐

We can use unit step functions to write any case-defined, up to the points where the discontinuity points of the unit step functions.

Example 5.21 *Express*

$$f(t) = \begin{cases} 0 & t < 1 \\ t^2 & 1 < t < 2 \\ -5 & 2 < t < 3 \\ \sin t & t > 3 \end{cases}$$

in terms of unit step functions.

Solution: We may rewrite this function as

$$f(t) = t^2[u(t-1) - u(t-2)] - 5[u(t-2) - u(t-3)] + (\sin t)u(t-3)$$

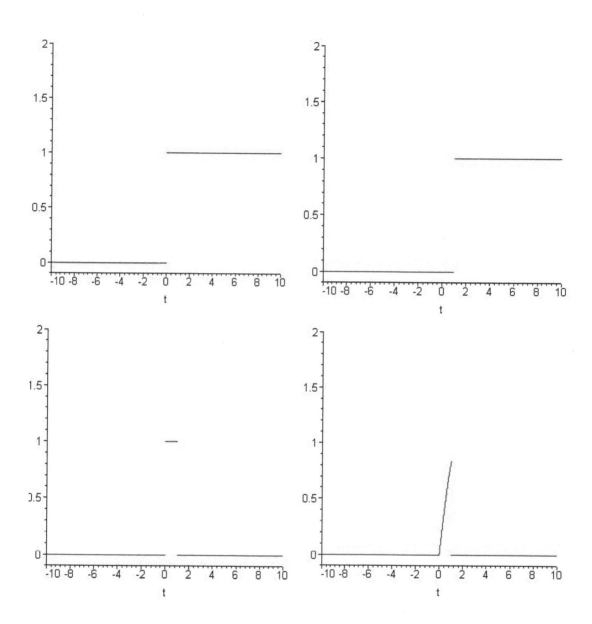

Figure 5.1: Plots of (a)-(d) in Exercise 5.20

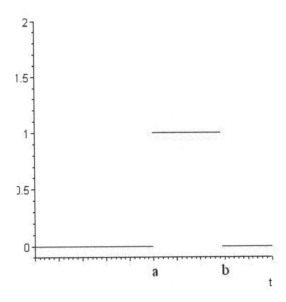

Figure 5.2: Plot of $u(t-a) - u(t-b)$, which is 1 on (a, b)

Note that this can be further simplified as

$$f(t) = t^2 u(t-1) - (5 + t^2)u(t-2) + (\sin t + 5)u(t-3)$$

$\square$

Below, we describe how to express a case defined function using unit step functions.

Expressing a Case-Defined Function

The function

$$f(t) = \begin{cases} f_1(t) & t_0 < t < t_1 \\ f_2(t) & t_1 < t < t_2 \\ \vdots & \vdots \\ f_n(t) & t_{n-1} < t < t_n \end{cases}$$

can be rewritten as

$$f(t) = f_1(t)[u(t - t_0) - u(t - t_1)] + f_2(t)[u(t - t_1) - u(t - t_2)] + \ldots$$

$$+ f_n(t)[u(t - t_{n-1}) - u(t - t_n)]$$

or

$$f(t) = \sum_{j=1}^{n} f_j(t)[u(t - t_{j-1}) - u(t - t_j)]$$

Note that if

$$f(t) = \begin{cases} f_1(t) & t_0 < t < t_1 \\ f_2(t) & t_1 < t < t_2 \\ \vdots & \vdots \\ f_n(t) & t_{n-1} < t \end{cases}$$

then we would express $f(t)$ as

$$f(t) = f_1(t)[u(t - t_0) - u(t - t_1)] + f_2(t)[u(t - t_1) - u(t - t_2)] + \ldots$$

$$+ f_{n-1}(t)[u(t - t_{n-2}) - u(t - t_{n-1})] + f_n(t)u(t - t_{n-1})$$

Laplace Transforms of Step Functions

Laplace Transform of $u(t - a)$

For $a \geq 0$,

$$\mathcal{L}[u(t - a)](s) = \frac{e^{-as}}{s}, \quad s > 0$$

More generally,

> Laplace Transform of $u(t-a)f(t-a)$ (Pre-Shift Theorem)
>
> For $a \geq 0$,
> $$\mathcal{L}[u(t-a)f(t-a)](s) = e^{-as}F(s) \quad (1)$$
> or
> $$\mathcal{L}[u(t-a)f(t)](s) = e^{-as}\mathcal{L}[f(t+a)] \quad (2)$$

Proof: By definition

$$\mathcal{L}[u(t-a)f(t-a)] = \int_0^\infty e^{-st}u(t-a)f(t-a)\, dt$$

Since $u(t-a) = 0$ for $t < a$, and $u(t-a) = 1$ for $t > a$, this integral becomes

$$\int_a^\infty e^{-st}f(t-a)\, dt.$$

Let $w = t - a$ and $dw = dt$. Then this integral becomes

$$\int_0^\infty e^{-s(w+a)}f(w)\, dw$$

or

$$e^{-sa}\int_0^\infty e^{-sw}f(w)\, dw = e^{-sa}\mathcal{L}[f(w)](s) = e^{-sa}\mathcal{L}[f(t)](s)$$

$$\square$$

Formula (1) above is useful when inverting while formula (2) is useful in the forward direction.

Example 5.22 *Find*

$$\mathcal{L}[u(t-7)t^2]$$

Solution: By formula (2) we have

$$\mathcal{L}[u(t-a)f(t)](s) = e^{-as}\mathcal{L}[f(t+a)]$$

where $a > 0$. So we have $f(t) = t^2$. To use the formula, we need $f(t+7)$, which means we replace all $t's$ by $(t+7)'s$ and expand. So

$$f(t+7) == (t+7)^2 = t^2 + 14t + 49.$$

Substituting:

$$\mathcal{L}[u(t-7)t^2] = e^{-7s}\mathcal{L}\left[t^2 + 14t + 49\right]$$

$$= e^{-7s}\left(\frac{2}{s^3} + \frac{14}{s^2} + \frac{49}{s}\right)$$

$\square$

Example 5.23 *Find*

$$\mathcal{L}[u(t-4)\sin 2t]$$

Solution: Again, we pre-shift to use the formula, where $a = 4$ and $f(t) = \sin 2t$
Shifting (and using a trig identity),

$$f(t+4) = \sin(2t+8) = \sin 2t \cos 8 + \cos 2t \sin 8$$

So:

$$\mathcal{L}[u(t-4)\sin 2t] == e^{-4s}\left(\cos 8 \left(\frac{2}{s^2+4}\right) + \sin 8 \left(\frac{s}{s^2+4}\right)\right)$$

$\square$

Inverse Laplace Transforms involving e^{-as} (Backward Pre-Shift Theorem)
For $a \geq 0$,

$$\mathcal{L}^{-1}[e^{-as}F(s)] = u(t-a)f(t-a),$$

where $F(s) = \mathcal{L}[f(t)](s)$.

Example 5.24 *Find*

$$\mathcal{L}^{-1}\left[e^{-4s}\frac{1}{s^4}\right]$$

Solution: We know for $F(s) = \frac{6}{s^4}$ that $f(t) = t^3$. So

$$\mathcal{L}^{-1}\left[e^{-4s}\frac{1}{s^4}\right] = \frac{1}{6}\mathcal{L}^{-1}\left[e^{-4s}\frac{6}{s^4}\right] = \frac{1}{6}u(t-4)(t-4)^3$$

$\square$

We now solve a differential equation arising from a spring mass system with discontinuous forcing.

Example 5.25 *Solve*

$$y'' + y = t^2 u(t-3), \quad y(0) = 0, \quad y'(0) = 0$$

Solution: Taking the Laplace transform of both sides and writing $\mathcal{L}[y(t)]$ as $Y(s)$, we obtain:

$$s^2 Y(s) + Y(s) = e^{-3s} \mathcal{L}\left[(t+3)^2\right] = e^{-3s} \mathcal{L}\left[t^2 + 6t + 9\right]$$

so

$$Y(s) = e^{-3s} \left[\frac{2}{s^3} + \frac{6}{s^2} + \frac{9}{s}\right] \frac{1}{s^2+1}$$

Onto partial fractions, we obtain

$$Y(s) = e^{-3s} \left[\frac{As}{s^2+1} + \frac{B}{s^2+1} + \frac{C}{s} + \frac{D}{s^2} + \frac{E}{s^3}\right]$$

We skip the partial fractions work and simply report that $A = -7$, $B = -6$, $C = 7$, $D = 6$, and $E = 2$.

So

$$Y(s) = e^{-3s} \left[\frac{-7s}{s^2+1} + \frac{-6}{s^2+1} + \frac{7}{s} + \frac{6}{s^2} + \frac{2}{s^3}\right].$$

Therefore

$$y(t) = u(t-3)\left[-7\cos(t-3) - 6\sin(t-3) + 7 + 6(t-3) + (t-3)^2\right]$$

Note that all instances of t must be replaced by $t - 3$ by the shift formula.

$\square$

Periodic Functions

Definition 5.26 *A function is **periodic** if for some $T > 0$, $f(t+T) = f(t)$ for all t. The smallest such positive value of T is called the **period** of $f(t)$.*

One way to define a periodic function is simply to specify its values on $[0, T]$ and then extend it. We define the **windowed version of a function** $f(t)$ to be

$$f_T(t) = \begin{cases} f(t) & 0 < t < T \\ 0 & else \end{cases}$$

or

$$f_T(t) = f(t)\left[u(t) - u(t - T)\right]$$

Then we can write:

$$f(t) = \sum_{k=-\infty}^{\infty} f_T(t - kT) = \sum_{k=-\infty}^{\infty} f(t - kT)\left[u(t - kT) - u(t - (k+1)T)\right],$$

but note that this function is not actually defined at the values of $t = 0, \pm, \pm 2T, ...,$ since the unit step functions are not defined there. Note that if we only only care about $f(t)$ when $t > 0$, then

$$f(t) = \sum_{k=0}^{\infty} f_T(t - kT) = \sum_{k=0}^{\infty} f(t - kT)\left[u(t - kT) - u(t - (k+1)T)\right].$$

Extending a Piece of a Function to a T-Periodic Function

Let $f(t)$ be a function defined for all t. The periodic extension of $f(t)$ via $f_T(t)$ is the function with period T given by

$$\widetilde{f}(t) = \sum_{k=0}^{\infty} f(t - kT)\left[u(t - kT) - u(t - (k+1)T)\right].$$

Note that this function is actually undefined for: $t = 0, T, 2T, 3T...$ This can be rewritten as:

$$\widetilde{f}(t) = f(t) + \sum_{k=1}^{\infty} \left[f(t - kT) - f(t - (k-1)T)\right] u(t - kT).$$

If we only care about this function on a finite interval, we do not need all the terms in this infinite sum.

Example 5.27 *Suppose that $f(t) = t$ and we want to create $f_T(t)$ for $T = 2$ and extend it to a periodic function $\widetilde{f}(t)$. Plot the graph of $\widetilde{f}(t)$ on $[0, 10]$ and express $\widetilde{f}(t)$ in terms of unite step functions on $[0, 10]$.*

Solution: Effectively, we are taking $f(t) = t$ on the interval $(0, 2)$ repeating it, so its graph on $[0, 10]$ is in Figure 5.4.

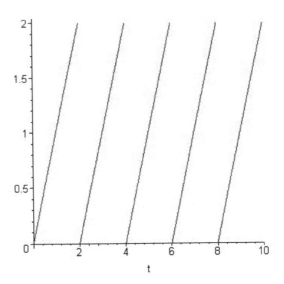

Figure 5.3: Plot of periodic function generated by $f(t) = t$ on $(0, 2)$

Note that for $t > 0$,

$$\widetilde{f}(t) = \sum_{k=0}^{\infty} (t - 2k) \left[u(t - 2k) - u(t - 2(k+1)) \right].$$

Note that this is (after expanding)

$$\widetilde{f}(t) = t - 2u(t - 2) - 2u(t - 4) - 2u(t - 6) - \dots$$

$$= t - 2 \sum_{k=1}^{\infty} u(t - 2k)$$

$\square$

Example 5.28 *Solve*

$$y'' + y = \widetilde{f}(t), y(0) = 0, \ y'(0) = 0$$

where $\widetilde{f}(t)$ is as in Example 5.27.

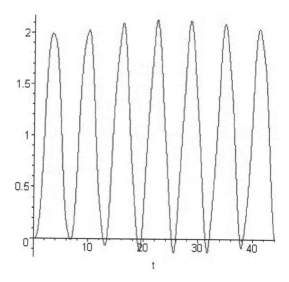

Figure 5.4: Solution of IVP in Example 5.28

Solution: Since

$$\widetilde{f}(t) = t - 2 \sum_{k=1}^{\infty} u(t - 2k)$$

we take the Laplace transform of both sides to obtain:

$$(s^2 + 1)Y(s) = \frac{1}{s^2} - 2 \sum_{k=1}^{\infty} \frac{e^{-2ks}}{s}$$

$$Y(s) = \frac{1}{s^2(s^2 + 1)} - 2 \sum_{k=1}^{\infty} \frac{e^{-2ks}}{s(s^2 + 1)}$$

$$Y(s) = \left(\frac{1}{s^2} - \frac{1}{s^2 + 1} \right) - 2 \left(\frac{1}{s} - \frac{s}{s^2 + 1} \right) \sum_{k=1}^{\infty} e^{-2ks}$$

so

$$y(t) = t - \sin(t) - 2 \sum_{k=1}^{\infty} (1 - \cos(t - 2k))u(t - 2k).$$

A plot of the solution for $t = 0$ to $t = 44$ is shown.

$\square$

The following is also helpful for a periodic function with windowed version $f_T(t)$.

Laplace Transform of Periodic Functions

For a periodic function $\widetilde{f}(t)$ with associated windowed version $f_T(t)$ we have

$$\mathcal{L}[\widetilde{f}(t)] = \frac{1}{1 - e^{-Ts}} F_T(s) = F_T(s) \sum_{k=0}^{\infty} e^{-kTs},$$

Proof: Since for $t > 0$ we have

$$f_T(t) = \widetilde{f}(t) \left[u(t) - u(t - T) \right]$$

Since $\widetilde{f}$ is $T-$periodc we have

$$f_T(t) = f(t)u(t) - f(t - T)u(t - T).$$

Taking the Laplace transform of both sides yields:

$$F_T(s) = \mathcal{L}[\widetilde{f}(t)] - e^{-sT}\mathcal{L}[\widetilde{f}(t)].$$

Therefore,

$$\mathcal{L}[\widetilde{f}(t)] = \frac{1}{1 - e^{-sT}} F_T(s).$$

Note that we have a the form of the sum of an infinite geometric sequence, namely:

$$\frac{1}{1 - e^{-sT}} = 1 + e^{-sT} + e^{-2sT} + \ldots$$

So

$$\mathcal{L}[\widetilde{f}(t)] = F_T(s) \sum_{k=0}^{\infty} e^{-kTs}.$$

$\square$

Exercises

In 1-5, write the function in terms of unit step functions and take the Laplace Transform

1. $f(t) = \begin{cases} 1 & t < 1 \\ e^t & t > 1 \end{cases}$

2.
$$f(t) = \begin{cases} \sin t & t < \pi \\ \cos t & t > \pi \end{cases}$$

3.
$$f(t) = \begin{cases} \sin(2t) & t < 2\pi \\ 0 & t > 2\pi \end{cases}$$

4.
$$f(t) = \begin{cases} 1 & 0 < t < 2 \\ 2 & 2 < t < 4 \\ 6 & t > 4 \end{cases}$$

5.
$$f(t) = \begin{cases} t^2 & 0 < t < 2 \\ 8 - t^2 & 2 < t < 5 \\ e^{-3t} & t > 5 \end{cases}$$

6. Solve $y'' + 2y' + 4y = u(t-2) - u(t-3), y(0) = 0, \; y'(0) = 0$.

7. Solve $y'' + 2y' + 4y = t^2 u(t-2) - t^2 u(t-3), y(0) = 0, \; y'(0) = 0$.

8. Solve $y'' + 2y' + 4y = e^t[u(t-2) - u(t-3)], y(0) = 0, \; y'(0) = 0$.

9. Graph the function $f(t) = 1 - u(t-1) + u(t-2) - u(t-3) + \dots$.

10. Solve $y'' + 2y' + 4y = f(t), y(0) = 0, \; y'(0) = 0$, where $f(t)$ is given in the previous problem.

11. Graph the function $f(t) = 2t - (2t-2)u(t-1) + (2t-4)u(t-2) - (2t-6)u(t-3) + \dots$.

12. Solve $y'' + 2y' + 4y = f(t), y(0) = 0, \; y'(0) = 0$, where $f(t)$ is given in the previous problem.

13. Consider $f(t) = e^{2t}$ made into a periodic function $\widetilde{f}(t)$ by taking $f_T(t)$ where $T = 1$.

   (a) Plot $\widetilde{f}(t)$ for $0 < t < 4$.

   (b) Find $\mathcal{L}[\widetilde{f}(t)]$

   (c) $y'' + 2y' + 3y = \widetilde{f}(t), y(0) = 0, \ y'(0) = 0,$

14. Use the differentiation theorem to verify that $\mathcal{L}[t \, u(t - a)] = e^{-as}\frac{1}{s^2}$

15. Use appropriate theorems to compute $\mathcal{L}[t \sin t e^t u(t - a)]$

5.5 Convolution and the Laplace Transform

We introduce a new operation $*$ between two functions called the **convolution**.

Convolution of Two Functions

Let $f(t)$ and $g(t)$ be two functions. Define a new function called the **convolution**, labeled as $f * g$ where:

$$(f * g)(t) = \int_0^t f(t - w)g(w) \ dw$$

Note that $f * g$ is itself a function of t. Moreover note that if we substitute $v = t - w$ then $dv = -dw$ and the integral becomes

$$\int_{w=t}^{w=0} f(v)g(t - v)(-1) \ dw = \int_{w=0}^{w=t} f(v)g(t - v) \ dw$$

which is $g * f$.

Example 5.29 *Find* $t * 1$

Solution: Since $f * g = g * f$, we can compute $1 * t$ easier, so let $f(t) = 1$ and $g(t) = t$. Then

$$f * g = \int_0^t f(t - w)g(w) \ dw = \int_0^t w \ dw = \frac{w^2}{2} \Big|_0^t = \frac{t^2}{2}$$

$\square$

We see that convolution is not the same as regular multiplication.

Example 5.30 *Find* $t * e^t$

Solution: We set $f(t) = t$ and $g(t) = e^t$.
  Then
$$f * g = \int_0^t f(t - w)g(w) \ dw = \int_0^t (t - w)e^w \ dw$$

$$= \int_0^t te^w - we^w \ dw$$

$$= (te^w - we^w + e^w))|_0^t = e^t - t - 1$$

□

The following result shows why convolutions are important:

Laplace Transform of the Convolution

Let $f(t)$ and $g(t)$ be two functions with Laplace transforms $F(s)$ and $G(s)$, respectively. Then:

$$\mathcal{L}[f * g] = F(s)G(s)$$

and

$$\mathcal{L}^{-1}[F(s)G(s)] = f * g$$

Example 5.31 *Find* $\mathcal{L}[t^2 * e^t]$

Solution: Since $\mathcal{L}[f * g] = F(s)G(s)$, we have

$$\mathcal{L}[f * g] = \frac{2}{s^3} \frac{1}{s - 1}$$

□

We can use convolution as an alternative to partial fractions as shown next.

Example 5.32 *Solve* $y'' + y = e^{2t}$, $y(0) = 0$, $y'(0) = 0$.

Solution: After taking Laplace transform of both sides we get:

$$(s^2 + 1)Y(s) = \frac{1}{s - 2}$$

or

$$Y(s) = \frac{1}{s^2 + 1} \frac{1}{s - 2}$$

so setting $F(s) = \frac{1}{s^2+1}$ and $G(s) = \frac{1}{s-2}$ we see that
$Y(s) = F(s)G(s)$ so $y(t) = f * g$ where $f(t) = \sin t$ and $g(t) = e^{2t}$.
The convolution is

$$\int_0^t e^{2(t-w)} \sin w \ dw$$

which is

$$\int_0^t e^{2(t-w)} \sin w \ dw.$$

This is the same thing as

$$e^{2t} \int_0^t e^{-2w} \sin w \ dw.$$

At this point, we could integrate by parts to get the solution, but we wish to introduce a slick trick to avoid integration by PARTS, since the integrand looks like the definition of a Laplace transform, where $s = 2$, and since $1 - u(w - t)$ is equal to zero for $w > t$ and is equal to one for $w < t$ (we view t as fixed), we may rewrite

$$e^{2t} \int_0^t e^{-2w} \sin w \ dw$$

$$= e^{2t} \int_0^t [1 - u(w - t)] e^{-2w} \sin w \ dw + e^{2t} \int_t^\infty [1 - u(w - t)] e^{-2w} \sin w \ dw$$

$$= e^{2t} \int_0^\infty [1 - u(w - t)] e^{-2w} \sin w \ dw$$

$$= e^{2t} \mathcal{L} \left[(1 - u(w - t)) \sin w \right] (2)$$

(note $s = 2$ in the Laplace transform definition).

$$= e^{2t} \left(\mathcal{L} \left[\sin w \right] (2) - \mathcal{L} [u(w - t) \sin w] (2) \right)$$

$$= e^{2t} \left(\mathcal{L} \left[\sin w \right] (2) - \mathcal{L} [u(w - t) \sin(w - t + t)] (2) \right)$$

$$= e^{2t} \left(\frac{1}{2^2 + 1} - \mathcal{L} \left[u(w - t) \sin(w - t) \cos t + u(w - t) \cos(w - t) \sin t \right] (2) \right)$$

$$= e^{2t} \left(\frac{1}{2^2 + 1} - \mathcal{L} \left[u(w - t) \sin(w - t) \cos t + u(w - t) \cos(w - t) \sin t \right] (2) \right)$$

$$= e^{2t}\left(\frac{1}{5} - \cos t \mathcal{L}[u(w-t)\sin(w-t)](2) - \sin t \mathcal{L}[u(w-t)\cos(w-t)](2)\right)$$

$$= e^{2t}\left(\frac{1}{5} - \cos t(e^{-2t}\frac{1}{2^2+1}) - \sin t(e^{-2t}\frac{2}{2^2+1})\right)$$

$$= \frac{1}{5}e^{2t} - \frac{1}{5}\cos t - \frac{2}{5}\sin t$$

Perhaps this was better done with PARTS, but we wanted to illustrate the power of the Laplace transform $\qquad\square$

The advantage of convolution is that we can solve **any** spring mass system without actually specifying a forcing function, as illustrated in the next example.

Example 5.33 *Find one solution to* $y'' + y = g(t)$ *for any* $g(t)$.

Solution: After taking Laplace transform of both sides we get:

$$(s^2 + 1)Y(s) = G(s)$$

or

$$Y(s) = \frac{1}{s^2+1}G(s)$$

so setting $F(s) = \frac{1}{s^2+1}$ and we see that
$Y(s) = F(s)G(s)$ so $y(t) = f * g$ where $f(t) = \sin t$.
The convolution (and hence one solution) is

$$y(t) = \int_0^t \sin(t-w)g(w) \; dw.$$

$\qquad\square$

<u>Derivative of a Convolution</u>
Consider $(f * g)(t)$, where f is continuous and g is differentiable.
Then
$$\frac{d}{dt}(f * g)(t) = (f') * g + f(0)g(t)$$

Proof: By definition we need to compute

$$\lim_{h \to 0} \frac{(f * g)(t + h) - (f * g)(t)}{h}$$

$$= \lim_{h \to 0} \frac{\int_0^{t+h} f(t + h - w)g(w) \, dw - \int_0^t f(t - w)g(w) \, dw}{h}$$

$$= \lim_{h \to 0} \frac{\int_0^t f(t + h - w)g(w) \, dw + \int_t^{t+h} f(t + h - w)g(w) \, dw - \int_0^t f(t - w)g(w) \, dw}{h}$$

$$= \lim_{h \to 0} \frac{\int_0^t \left(f(t + h - w) - f(t - w) \right) g(w) \, dw + \int_t^{t+h} f(t + h - w)g(w) \, dw}{h}$$

$$= \int_0^t \lim_{h \to 0} \left(\frac{f(t + h - w) - f(t - w)}{h} \right) g(w) \, dw + \lim_{h \to 0} \frac{\int_t^{t+h} f(t + h - w)g(w) \, dw}{h}$$

$$= \int_0^t f'(t - w)g(w) \, dw + \lim_{h \to 0} \frac{\int_t^{t+h} f(t + h - w)g(w) \, dw}{h}$$

Note that to evaluate the second limit, we note by the Mean Value theorem for integrals that there is a c value inside $[t, t+h]$ so that $\int_t^{t+h} f(t+h-w)g(w) \, dw = hf(t + h - c)g(c)$. So the second limit becomes (by continuity):

$$\lim_{h \to 0} \frac{hf(t + h - c)g(c)}{h} = f(0)g(t)$$

so

$$(f * g)' = (f') * g + f(0)g(t)$$

$\square$

Note: Similarly, $(f * g)' = f * (g') + f(t)g(0)$

Convolution and Second Order Linear with Constant Coefficients

Consider

$$ay'' + by' + cy = g(t), \quad y(0) = 0, \quad y'(0) - 0.$$

The solution is $y(t) = f * g$ where $F(s) = \frac{1}{as^2+bs+c}$, which is called the **transfer function** and we call $f(t)$ **impulse response function** for this second order DE.

By superposition, we obtain the following:

Convolution and Second Order Linear with Constant Coefficients

Consider
$$ay'' + by' + cy = g(t), \ y(0) = c_1, \ y'(0) = c_2.$$

If we have the particular solution to the homogeneous $y_{homo\ part}(t)$ that satisfies the initial conditions $y(0) = c_1$ and $y'(0) = c_2$ then

$$y(t) = y_{homo\ part}(t) + f * g(t)$$

will solve the nonhomogeneous IVP, where $F(s) = \frac{1}{as^2+bs+c}$.

Exercises

In 1-5, find the convolution.

1. $t^2 * t^3$

2. $e^t * e^{3t}$

3. $\cos t * 1$

4. $\cos t * \sin t$

5. $u(t-1) * 1$

6. Use convolution to solve $y'' + y = 2, \quad y(0) = 0, y'(0) = 0$

7. Use convolution to solve $y'' - 4y = t, \quad y(0) = 0, y'(0) = 0$

8. For a fixed constant β, use convolution to solve $y'' + 4y = \sin \beta t, \quad y(0) = 0, y'(0) = 0$

9. Use an integral approximation to estimate $y(1)$ for $y'' + 4y = e^{t^2}, \quad y(0) = 0, y'(0) = 0$

10. Find the impulse response function for an overdamped spring mass system $my'' + by' + ky = g(t)$

11. Find the impulse response function for an underdamped spring mass system $my'' + by' + ky = g(t)$

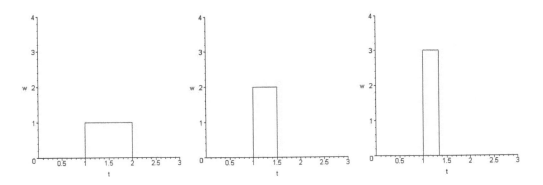

Figure 5.5: $w_1(t)$, $w_2(t)$, and $w_3(t)$.

5.6 δ and Γ Functions

Dirac δ-Functions

We introduce the notion of the Dirac δ function, which is the limit of a sequence of step functions. We build the δ function at $t = 1$.

Consider the following sequence of functions:

$$w_n(t) = n[u(t-1) - u(t - (1 + \frac{1}{n}))]$$

Note that for any integer n,

$$\int_0^\infty w_n(t) \, dt = \int_0^\infty n[u(t-1) - u(t - (1 + \frac{1}{n}))] \, dt = 1.$$

A figure showing $w_1(t)$, $w_2(t)$, and $w_3(t)$ is included.

The limit of these functions is called the Dirac delta function $\delta(t-1)$. This not actually a function, since it is infinite or undefined at $t = 1$. Clearly, it is zero for all $t \neq 1$. Also, for $a < 1 < b$, this object will satisfy $\int_a^b \delta(t-1) \, dt = 1$. (Technically, such objects are called distributions in mathematics).

In general we will define the Dirac functions a bit differently than we did above, in terms of the base of the rectangles which go to zero, instead of the height which goes to infinity.

$$\boxed{\begin{array}{c} \underline{\text{Dirac } \delta \text{ Function}} \\[2mm] \delta(t-a) = \lim_{b \to 0} \frac{1}{b}[u(t-a) - u(t-(a+b))] \end{array}}$$

$$\boxed{\begin{array}{l} \text{For any } a > 0 \qquad \begin{array}{c} \underline{\text{Laplace Transform of } \delta(t-a)} \\[2mm] \mathcal{L}[\delta(t-a)] = e^{-as} \end{array} \end{array}}$$

Proof: Since

$$\int_0^\infty e^{-st} \delta(t-a) \, dt = \int_0^\infty e^{-st} \lim_{b \to 0} \frac{1}{b}[u(t-a) - u(t-(a+b))] \, dt$$

$$\lim_{b \to 0} \frac{1}{b} \int_a^{a+b} e^{-st} \, dt$$

$$\lim_{b \to 0} \frac{1}{b} \left(\frac{e^{-st}}{-s} \right) \Big|_a^{a+b}$$

$$\lim_{b \to 0} \frac{1}{bs}[e^{-sa} - e^{-sa-sb}]$$

$$e^{-sa} \lim_{b \to 0} \frac{1}{bs}[1 - e^{-sb}]$$

Applying L'Hopital's Rule,

$$c^{-sa} \lim_{b \to 0} \frac{1}{s}[se^{-sb}] = e^{-sa}$$

$\square$

The main application of the delta function is that it resembles a finite impulse being applied to a system instantaneously, much like hitting a spring/mass system with a hammer.

Example 5.34 *Solve* $y'' + 2y' + y = \delta(t-4)$, $y(0) = 0$, $y'(0) = 0$.

Solution: Taking the Laplace transform of both sides yields $(s^2+2s+1)Y(s) = e^{-4s}$ so $Y(s) = \frac{1}{(s+1)^2}e^{-4s}$, therefore

$$y(t) = e^{-(t+4)}u(t-4)(t-4)$$

$\square$

Convolution with $\delta(t-a)$
For any $a > 0$ and suppose $\overline{f(t-a)}$ is continuous at $t = a$ (i.e. $f(t)$ is continuous at $t = 0$), then $$\delta(t-a) * f(t) = u(t-a)f(t-a)$$

Proof: Suppose that $t < a$, then

$$\delta(t-a) * f(t) = \int_0^t f(t-w)\lim_{b\to 0}\frac{1}{b}[u(w-a) - u(w-(a+b))]\ dw = 0$$

Suppose that $t > a$

$$\delta(t-a) * f(t) = \int_0^t f(t-w)\lim_{b\to 0}\frac{1}{b}[u(w-a) - u(w-(a+b))]\ dw$$

$$= \lim_{b\to 0}\int_a^{a+b}\frac{1}{b}f(t-w)\ dw$$

By the mean value theorem for integrals applied to $f(t-a)$ on the interval $[a, a+b]$, $\int_a^{a+b}\frac{1}{b}f(t-w)\ dw$ is the average value of $f(t-w)$ on $[a, a+b]$. Let $f(t-z_b)$ be the average value which is achieved at some value z_b in $[a, a+b]$. By continuity, $\lim_{b\to 0} z_b = a$ and since $f(t-a)$ is continuous at a we have $\lim_{b\to 0} f(t-z_b) = f(t-a)$, so

$$\lim_{b\to 0}\int_a^{a+b}\frac{1}{b}f(t-w)\ dw = f(t-a)$$

Since for $t < a$ we have $\delta(t-a)*f(t) = 0$ and for $t > a$ we have $\delta(t-a)*f(t) = f(t-a)$ then $\delta(t-a) * f(t) = u(t-a)f(t-a)$. $\square$

The next example illustrates what happens if a spring/mass system is repeatedly forced by a hammer-like impulse.

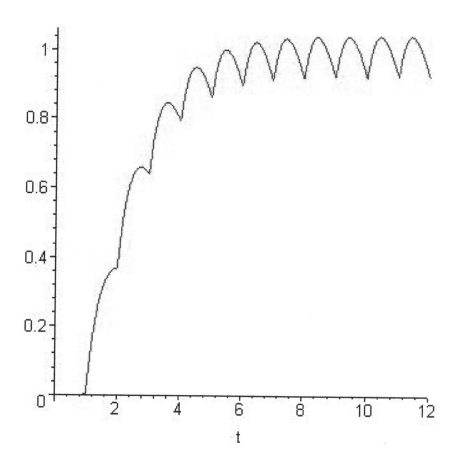

Figure 5.6: A spring/mass system hit by a hammer repeatedly.

Example 5.35 *Solve* $y'' + 2y' + y = \sum_{k=1}^{\infty} \delta(t-k)$, $y(0) = 0$, $y'(0) = 0$. *Plot the solution for $0 < t < 12$.*

Solution: Taking the Laplace transform of both sides yields $(s^2 + 2s + 1)Y(s) = \sum_{k=1}^{\infty} e^{-sk}$ so $Y(s) = \frac{1}{(s+1)^2} \sum_{k=1}^{\infty} e^{-sk}$, therefore

$$y(t) = \sum_{k=1}^{\infty} e^{-(t-k)} u(t-k)(t-k)$$

□

The Γ-Function

We define the Gamma (or Γ) function which is a useful extension of the notion of a factorial.

The Gamma Function $\Gamma(x)$

The Gamma function is defined to as

$$\Gamma(x) = \int_0^\infty t^{x-1}e^{-t}\, dt.$$

Note that the Gamma function is of the form of a Laplace transform, namely it is $\mathcal{L}[t^{x-1}](1)$ for any x. In particular, when x is a positive integer, $\mathcal{L}[t^{x-1}](1) = (x-1)!$. So $\Gamma(x+1)$ is an extension of $n!$.

It turns out that for any x we have $\Gamma(x+2) = (x+1)\Gamma(x+1)$ similar to the factorial where $(n+1)! = (n+1)n!$.

Theorem 5.36 *For any $x \geq 0$ we have $\Gamma(x+1) = x\Gamma(x)$*

Proof: By definition,

$$\Gamma(x+1) = \int_0^\infty t^x e^{-t}\, dt.$$

Using integration by parts with $u = t^x$ and $dv = e^{-t}\, dt$ we have $du = xt^{x-1}dx$ and $v = -e^{-t}$. So we get

$$= -\left(t^x e^{-t}\right)\big|_0^\infty + \int_0^\infty xt^{x-1}e^{-t}\, dt$$

$$= x\Gamma(x)$$

$\square$

The Gamma Function and $\mathcal{L}[t^x](s)$

For fixed x,

$$\mathcal{L}[t^x](s) = \frac{\Gamma[x+1]}{s^{x+1}}$$

Proof:

By definition,

$$\mathcal{L}[t^x](s) = \int_0^\infty t^x e^{-st} \, dt = \int_0^\infty t^x e^{-st} \, dt$$

Let $v = st$, then integral is

$$\int_0^\infty t^x e^{-st} \, dt = \int_0^\infty \left(\frac{v}{s}\right)^x e^{-v} \frac{1}{s} \, dv$$

$$\int_0^\infty t^x e^{-st} \, dt = \frac{1}{s^{x+1}} \int_0^\infty v^x e^{-v} \, dv$$

$$\int_0^\infty t^x e^{-st} \, dt = \frac{\Gamma(x+1)}{s^{x+1}}$$

☐

It turns out that

$$\Gamma(\tfrac{1}{2}) = \sqrt{\pi}$$

Exercises

1. $y'' + 2y' + y = \delta(t-1) + \delta(t-2) + \delta(t-3)$, $y(0) = 0$, $y'(0) = 0$. Plot the solution for $0 < t < 5$.

2. $y'' + 2y' + y = \sqrt[3]{t}$, $y(0) = 0$, $y'(0) = 0$. (Express as a convolution and plot the solution for $0 < t < 1$).

3. $y'' + y = \Gamma(t+1)$, $y(0) = 0$, $y'(0) = 0$. (Be careful with the names of variables! Express as a convolution and plot the solution for $0 < t < 4$).

Chapter 6

Series Solutions

6.1 Review of Power Series

In this section we will analyze power series and Taylor series solutions to differential equations. We remind the reader that not all power series converge. Therefore, one might obtain a power series solution and have it not converge. The reader should be able to apply convergence results from second semester calculus to determine convergence.

We remind the reader of several definitions and results, which we will state without proof.

Power Series

A **power serics** is centered at x_0 is a series of the form

$$a_0 + a_1(x - x_0) + a_2(x - x_0)^2 + a_3(x - x_0)^3 + \ldots$$

which can be compressed as

$$\sum_{j=0}^{\infty} a_j(x - x_0)^j.$$

When the center x_0 is zero, we call the power series a Maclaurin Series.

Power Series-Interval and Radius of Convergence

For any power series,

$$\sum_{j=0}^{\infty} a_j (x - x_0)^j$$

there exists an $R \geq 0$ so that the series converges for any x in the interval $(x_0 - R, x_0 + R)$ and diverges for all x that are outside the interval and not endpoints of the interval.

The interval $(x_0 - R, x_0 + R)$ is called the interval of convergence and the value R is called the radius of convergence.

If $R = +\infty$ then the power series converges for all x, and if $R = 0$ then the power series converges only at $x = x_0$.

The series may converge or diverge at the endpoints, further analysis is required at these points.

Power Series-Derivatives

For any power series,

$$g(x) = \sum_{j=0}^{\infty} a_j (x - x_0)^j$$

with radius of convergence R

The *term-by-term derivative* of the power series:

$$\sum_{j=0}^{\infty} j a_j (x - x_0)^{j+1}$$

also has radius of convergence R, moreover

$$g'(x) = \sum_{j=0}^{\infty} j a_j (x - x_0)^{j-1}$$

Power Series-Integrals

For any power series,

$$g(x) = \sum_{j=0}^{\infty} a_j (x - x_0)^j$$

with radius of convergence R the *term-by-term integral* of the original power series:

$$C + \sum_{j=0}^{\infty} \frac{a_j}{j+1} (x - x_0)^{j+1}$$

also has radius of convergence R, moreover

$$\frac{d}{dx} \left(C + \sum_{j=0}^{\infty} \frac{a_j}{j+1} (x - x_0)^{j+1} \right) = g(x)$$

This result seems more trivial than it is. The derivative of finite sum of terms is clearly the sum of the derivatives, but this is not necessarily true for an infinite sum. The above result says that the derivative of the sum is the sum of the derivatives for a power series on its interval of convergence.

The next result is one of the most powerful results from calculus.

Taylor Series

For any function $f(x)$ Suppose that for all z in $(x_0 - R, x_0 + R)$

$$\frac{f^{(k)}(z)}{k!} (x - x_0)^k$$

limits to 0 at k goes to infinity, then the Taylor series:

$$f(x_0) + f'(x_0)(x - x_0) + \frac{1}{2} f''(x_0)(x - x_0)^2 + \frac{1}{3!} f^{(3)}(x_0)(x - x_0)^3 + \dots$$

converges on $(x_0 - R, x_0 + R)$ and the limits to $f(x)$, so

$$f(x) = \sum_{j=0}^{\infty} \frac{1}{j!} f^{(j)}(x_0)(x - x_0)^j$$

for x in $(x_0 - R, x_0 + R)$.

One should be careful to realize that not all Taylor series converge to the functions that gave rise to them. For $f(x) = \begin{cases} e^{-\frac{1}{x^2}} & x \neq 0 \\ 0 & x = 0 \end{cases}$ has $f^{(k)}(0) = 0$ for all k which means the associated Taylor series centered at zero is zero, which is clearly not equal to the function.

6.2   Power Series Solutions

In this section we assume that a power series solution to a differential equation exists. We then will be faced with a giant coefficient match in order to find the appropriate coefficients of the power series solution. Note that we are making an assumption here that a power series solution exists and that t is within the power series interval of convergence. (Here we plan to use the variable t instead of x.) In other words, we assume the solution $y(t)$ has the form $y(t) = a_0 + a_1(t - t_0) + a_2(t - t_0)^2 + \ldots.$

This brings us to the 'dot-dot-dot' notation. A student should only be allowed to write 'dot-dot-dot' if they can easily generate the next term (and any other term). For example, the next term in $y(t) = a_0 + a_1 t + a_2 t^2 + \ldots.$ is clearly $a_3 t^3$ and the term involving t^n would be $a_n t^n$.

Next, we take derivatives term-by-term and match coefficients by plugging into the differential equation. The goal is to solve for all of the coefficients a_n. Note that this is possible when initial conditions are given. One can also find general solutions using this method by leaving n constants free.

Example 6.1 *Find the first five terms in the Maclaurin series solution to*

$$y'' + 4y = 0, \quad y(0) = 1, \ y'(0) = 3.$$

Explain how to generate the full series.

Solution: Here we assume that $y(t)$ can be expressed as a power series (with variable t).

So

$$y(t) = a_0 + a_1 t + a_2 t^2 + \ldots.$$

Since $y(0) = 1$, we see that $a_0 = 1$
Clearly,

$$y'(t) = a_1 + 2a_2 t + 3a_3 t^2 + \ldots$$

and since $y'(0) = 3$ we have $a_1 = 3$.

$$y''(t) = 2a_2 + 6a_3 t + 12a_4 t^2 + \dots$$

Note that next term would be $24a_5 t^3$ and that the t^n term would be $(n+2)(n+1)a_{n+2}t^n$.

We now plug this info into the differential equation.

We obtain:

$$\left(2a_2 + 6a_3 t + 12a_4 t^2 + \dots\right) + 4\left(1 + 3t + a_2 t^2 + \dots\right) = 0$$

We now compute coefficients of the power series that results on the left-hand side.

The constant term is: $2a_2 + 4$

The coefficient of t is: $6a_3 + 12$

The coefficient of t^2 is: $12a_4 + 4a_2$

The coefficient of t^n is: $(n+1)(n+2)a_{n+2} + 4a_n$

The right hand side of the DE is zero, so all of these coefficients must be set equal to zero.

We obtain:

$2a_2 + 4 = 0$ so $a_2 = -2$

$6a_3 + 12 = 0$ so $a_3 = -2$

$12a_4 + 4a_2 = 0$ so $12a_4 = 8$ and $a_4 = \frac{2}{3}$

Using the formula $(n+1)(n+2)a_{n+2} + 4a_n = 0$ we could obtain any coefficient. Therefore the solution is given by

$$y(t) = 3 + t - 2t^2 - 2t^3 + \frac{2}{3}t^4 + \dots$$

where the remaining coefficients could be determined using the above formula.

□

This method requires all terms in the differential equation to be converted to power series with the appropriate center. Taylor's Theorem can be useful for this.

Example 6.2 *Find the first five terms in the Maclaurin series solution to*

$$y'' + ty = e^{-t}.$$

Explain how to generate the full series.

Solution: Here we assume that $y(t)$ can be expressed as a power series (with variable t).

So
$$y(t) = a_0 + a_1 t + a_2 t^2 + \ldots.$$

Next, we note that $ty(t)$ has power series

$$ty(t) = a_0 t + a_1 t^2 + a_2 t^3 + \ldots.$$

and that $e^{-t} = 1 - t + \frac{1}{2} t^2 - \frac{1}{3!} t^3 + \ldots$

$$y''(t) = 2a_2 + 6a_3 t + 12 a_4 t^2 + \ldots$$

Writing the DE we have
$$y'' + ty = e^{-t}$$

Plugging in, we obtain:

$$\left(2a_2 + 6a_3 t + 12 a_4 t^2 + \ldots \right) + \left(a_0 t + a_1 t^2 + a_2 t^3 + \ldots \right) = 1 - t + \frac{1}{2} t^2 - \frac{1}{3!} t^3 + \ldots$$

We match coefficients:
The constant term: $2a_2 = 1$
The t term: $6a_3 + a_0 = -1$
The t^2 term: $12 a_4 + a_1 = \frac{1}{2}$
The t^3 term: $20 a_5 + a_2 = \frac{1}{6}$
The t^n term: $(n+1)(n+2) a_{n+2} + a_{n-1} = (-1)^n \frac{1}{n!}$
The idea here is that we must leave a_0 and a_1 free, since no initial conditions are given. All other terms can be computed in terms of a_0 and a_1.
$a_2 = \frac{1}{2}$
$a_3 = -\frac{1}{6}(1 + a_0)$
$a_4 = \frac{1}{12}(\frac{1}{2} - a_1)$
Therefore $y(t) = a_0 + a_1 t + \frac{1}{2} t^2 - \frac{1}{6}(1 + a_0) t^3 + \frac{1}{12}(\frac{1}{2} - a_1) t^4 + \ldots$
We can generate as many terms as we like in terms of a_0 and a_1 by using above formula derived for the t^n term to derive further quantities.

In particular a_6 can be computed taking $n = 4$ in

$$(n+1)(n+2) a_{n+2} + a_{n-1} = (-1)^n \frac{1}{n!}$$

or

$$(4+1)(4+2)a_{4+2} + a_{4-1} = (-1)^4 \frac{1}{4!}$$

we get

$$a_6 = \frac{1}{30}(\frac{1}{24} - a_3)$$

or

$$a_6 = \frac{1}{30}(\frac{1}{24} + \frac{1}{6}(1 + a_0)).$$

Again all coefficients can theoretically be solved in terms of a_0 and a_1. $\square$

The next example uses the fact that power series can be 'infinitely' multiplied just as polynomials can be multiplied.

Power Series-Multiplication

For any power series,

$$f(x) = \sum_{j=0}^{\infty} a_j(x - x_0)^j$$

with radius of convergence R
and

$$g(x) = \sum_{j=0}^{\infty} b_j(x - x_0)^j$$

with radius of convergence R
The *product* of the two power series has radius of convergence R:

$$\left(a_0 + a_1(x - x_0) + a_2(x - x_0)^2 + ...\right)\left(b_0 + b_1(x - x_0) + b_2(x - x_0)^2 + ...\right) :$$

has constant term $a_0 b_0$
$(x - x_0)$ has coefficient $a_1 b_0 + a_0 b_1$
$(x - x_0)^2$ has coefficient $a_2 b_0 + a_1 b_1 + a_0 b_2$
$(x - x_0)^n$ has coefficient $a_n b_0 + a_{n-1} b_1 + ... + a_1 b_{n-1} + a_0 b_n$.

Example 6.3 *Find the first six terms in the Maclaurin series solution to*

$$y'' + e^t y = 1$$

in terms of a_0 and a_1.

Solution: As before, assume that

$$y(t) = a_0 + a_1 t + a_2 t^2 + \ldots.$$

then

$$y''(t) = 2a_2 + 6a_3 t + 12a_4 t^2 + \ldots$$

recall that $\cos t = 1 - \frac{1}{2} t^2 + \frac{1}{4!} t^4 - \frac{1}{6!} t^6 + \ldots$

Plugging into the DE:

$$\left(2a_2 + 6a_3 t + 12a_4 t^2 + \ldots\right) + \left(1 - \frac{1}{2} t^2 + \frac{1}{4!} t^4 - \frac{1}{6!} t^6 + \ldots\right)\left(a_0 + a_1 t + a_2 t^2 + \ldots\right) = 1$$

Matching coefficients:

The constant term is $2a_2 + a_0 = 1$

The coefficient of t is $6a_3 + a_1 = 0$

The coefficient of t^2 is $12a_4 + a_2 - \frac{1}{2} a_0 = 0$

The coefficient of t^3 is $20a_5 + a_3 - \frac{1}{2} a_1 = 0$

We obtain $a_2 = \frac{1}{2}(1 - a_0)$

$a_3 = -\frac{1}{6}(a_1)$

$a_4 = \frac{1}{12}(\frac{1}{2} a_0 - a_2) = \frac{1}{12}(\frac{1}{2} a_0 - \frac{1}{2}(1 - a_0)) = \frac{1}{12}(a_0 - \frac{1}{2})$

$a_5 = \frac{1}{20}(\frac{1}{2} a_1 - a_3) = \frac{1}{20}(\frac{1}{2} a_1 + \frac{1}{6}(a_1)) = \frac{1}{20}(\frac{2}{3} a_1) = \frac{3}{10} a_1.$

So

$$y(t) = a_0 + a_1 t + \frac{1}{2}(1 - a_0)t^2 - \frac{1}{6}(a_1)t^3 + \frac{1}{12}(a_0 - \frac{1}{2})t^4 + \frac{3}{10} a_1 t^5 + \ldots$$

$\square$

Example 6.4 *Find the first four non-zero terms of a power series solution to centered at $t = 1$ to the DE*

$$y'' + ty = \frac{1}{t} \quad y(1) = 1, y'(1) = -2$$

Solution: As before, assume that

$$y(t) = a_0 + a_1(t - 1) + a_2(t - 1)^2 + \ldots.$$

and

$$y''(t) = 2a_2 + 6a_3(t-1) + 12a_4(t-1)^2 + 20a_5(t-1)^3....$$

We immediately have $a_0 = 1$ and $a_1 = -2$.

In order to match coefficients in the DE, we need all power series in the DE must be centered at 1, we write $t = (t-1) + 1$. We also expand $\frac{1}{t} = \frac{1}{1-[-(t-1)]}$ which allows us to write it as a geometric series with ratio $-(t-1)$ so

$$\frac{1}{t} = 1 - (t-1) + (t-1)^2 - (t-1)^3 + (t-1)^4 - ...$$

(This could have also been obtained by Taylor's Theorem).

So our DE becomes

$$y'' + (t-1)y + y = 1 - (t-1) + (t-1)^2 - (t-1)^3 + (t-1)^4 - ...$$

or

$$\left(2a_2 + 6a_3(t-1) + 12a_4(t-1)^2 + ...\right)$$

$$+ \left(a_0(t-1) + a_1(t-1)^2 + a_2(t-1)^3 + ...\right)$$

$$+ \left(a_0 + a_1(t-1) + a_2(t-1)^2 + ...\right) =$$
$$1 - (t-1) + (t-1)^2 - (t-1)^3 + (t-1)^4 - ...$$

The constant term yields $2a_2 + a_0 = 1$
The coefficient of $(t-1)$ yields $6a_3 + a_0 + a_1 = -1$
The coefficient of $(t-1)^2$ yields $12a_4 + a_1 + a_2 = 1$
The coefficient of $(t-1)^3$ is $20a_5 + a_3 + a_2 = -1$
We obtain $a_2 = 0$,
$6a_3 + 1 - 2 = -1$ so $a_3 = 0$,
$12a_4 - 2 = 1$ so $a_4 = -\frac{1}{4}$,
and, $20a_5 = -1$ so $a_5 = -\frac{1}{20}$.
Thus the power series solution (if it exists) has the form

$$y(t) = 1 - 2(t-1) - \frac{1}{6}(t-1)^4 - \frac{1}{20}(t-1)^5....$$

□

Exercises

In 1-4, find the first 5 non-zero terms of a Maclaurin Series solution to the IVP.

1. $y'' + ty' + ty = 0$, $y(0) = 1$, $y'(0) = 2$.

2. $y'' + ty = \sin t$, $y(0) = 1$, $y'(0) = 0$.

3. $y^{(3)} + ty = 0$, $y(0) = 1$, $y'(0) = -1$, $y''(0) = 2$.

4. $y'' + t^3 y = t \sin t$, $y(0) = 1$, $y'(0) = 1$.

In 5-8, find the first 5 non-zero terms of a Maclaurin Series solution of a general solution in terms of a_0 and a_1.

5. $y'' + ty' = e^t$.

6. $y'' + ty = \cos t$

7. $y'' + t^2 y' + ty = 2$

8. $y'' + y' + t^4 y = t - 2t + 4$

Find the first 5 non-zero terms of a Maclaurin Series solution of a general solution in terms of a_0 and a_1.

9. $y'' + e^t y = 0$

Find the first 5 non-zero terms of a general power series solution centered at $t = 1$ for the DE

10. $y'' + y = \ln t$

6.3 Taylor Series Methods

Sometimes it is easier to simply differentiate the DE to obtain the coefficients of the Taylor series.

Example 6.5 *Use Taylor's Theorem to find the first four non-zero terms of a power series solution centered at $t = 1$ to the DE*

$$y'' + ty = \frac{1}{t} \quad y(1) = 1, y'(1) = -2$$

Solution: We assume that the solution has a Taylor series representation

$$y(t) = y(1) + y'(1)(t - 1) + \frac{y''(1)}{2!}(t - 1)^2 + \frac{y^{(3)}(1)}{3!}(t - 1)^3 + \dots$$

We have $y(1)$ and $y'(1)$ already. To obtain $y''(1)$ we simply consult the DE when $t = 1$

$$y''(1) + 1y(1) = \frac{1}{1}$$

so $y''(1) = 0$

To obtain $y^{(3)}(1)$ we first differentiate the DE with respect to t

$$\frac{d}{dt}(y'' + ty) = \frac{d}{dt}\left(\frac{1}{t}\right)$$

so

$$y^{(3)} + ty' + y = -t^{-2}$$

plugging in $t = 1$, we see

$$y^{(3)}(1) + 1y'(1) + y(1) = -1^{-2}$$

so

$$y^{(3)}(1) = 0$$

Continuing, we derive a differential equation to obtain $y^{(4)}(1)$

$$\frac{d}{dt}\left(y^{(3)} + ty' + y\right) = \frac{d}{dt}\left(-t^{-2}\right)$$

so

$$y^{(4)} + ty'' + 2y' = 2t^{-3}$$

evaluating at $t = 1$ we obtain

$$y^{(4)}(1) + 1y''(1) + 2y'(1) = 21^{-3}$$

or

$$y^{(4)}(1) = 6$$

Similarly, we derive

$$y^{(5)} + ty^{(3)} + 3y'' = -6t^{-4}$$

to obtain $y^{(5)}(1) = -6$

In all, up to five non-zero terms

$$y(t) = y(1) + y'(1)(t - 1) + \frac{y''(1)}{2!}(t - 1)^2 + \frac{y^{(3)}(1)}{3!}(t - 1)^3 + \ldots$$

$$y(t) = 1 - 2(t - 1) - \frac{1}{4}(t - 1)^4 - \frac{1}{20}(t - 1)^5$$

This is the same series we obtained in the previous section. $\square$

We note that one can simply truncate a Taylor series solution to obtain an approximation to solutions. Combined with the Taylor Error Theorem, this can be a powerful tool for approximating solutions.

Undamped Pendulum

The equation governing the angle of an undamped pendulum is modeled by the DE

$$\theta'' + \frac{g}{L}\sin\theta = 0$$

, where $\theta = 0$ corresponds to the vertical position, where g is acceleration due to gravity and L is the length of the pendulum to the fixed point about which it is turning. The equation is independent of mass. See Figure 6.1

To simplify matters we will assume that $\frac{g}{L} = 1$. This is actually not as much of an assumption as one might guess, since there is a change of variables (which changes the units that time is measures in to) that accomplishes this namely let $\tau = \frac{\sqrt{g}}{\sqrt{L}}t$ then $\frac{d\theta}{d\tau} = \frac{d\theta}{dt}\frac{dt}{d\tau} = \frac{d\theta}{dt}\frac{\sqrt{L}}{\sqrt{g}}$

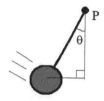

Figure 6.1: A Pendulum

Similarly,

$$\frac{d}{d\tau}\left(\frac{d\theta}{d\tau}\right) = \frac{d}{d\tau}\left(\frac{d\theta}{dt}\frac{\sqrt{L}}{\sqrt{g}}\right) = \frac{\sqrt{L}}{\sqrt{g}}\left(\frac{d}{d\tau}\left(\frac{d\theta}{dt}\right)\right) = \frac{\sqrt{L}}{\sqrt{g}}\left(\frac{d^2\theta}{dt^2}\right)\frac{dt}{d\tau} = \frac{L}{g}\frac{d^2\theta}{dt^2}$$

$$= -\sin\theta$$

So for time measured in τ units

$$\theta'' + \sin\theta = 0$$

This differential equation cannot be solved explicitly, but for any initial conditions, one can obtain a Taylor series solution and plot it.

Example 6.6 *(a) Use Taylor's Theorem to find the first three non-zero terms of a Maclaurin series solution to the DE/IVP*

$$\theta'' + \sin\theta = 0, \quad \theta(0) = 0, \quad \theta'(0) = 1$$

Solution: We have

$$\theta'' + \sin\theta = 0$$

so for $\theta(0) = 0$ we have $\theta''(0) = 0$.

Differentiating, we obtain

$$\theta^{(3)} + (\cos\theta)\theta' = 0$$

which implies that $\theta^{(3)}(0) = -1$.

Differentiating again, we obtain

$$\theta^{(4)} + (\cos\theta)\theta'' - (\sin\theta)(\theta')^2 = 0$$

which implies that $\theta^{(4)}(0) = 0$.

Differentiating again, we obtain

$$\theta^{(5)} + (\cos\theta)\theta^{(3)} - 3(\sin\theta)(\theta')(\theta'') - \cos\theta(\theta')^3 = 0$$

which implies that $\theta^{(5)}(0) = 2$.

Therefore $\theta(t) \approx t - \frac{t^3}{6} + 2\frac{t^5}{5!}$. $\square$

Exercises

In 1-2, find the first 5 non-zero terms of a Taylor series solution to the IVP centered at $t = 0$.

1. $y'' + e^y = 0$, $\quad y(0) = 0$, $y'(0) = 1$.

2. $y'' + \cos y = t$, $\quad y(0) = 0$, $y'(0) = 1$.

In 3-4, find the first 5 non-zero terms of a Taylor series solution to the IVP centered at the indicated center.

3. $y'' + \ln y = \dfrac{1}{t}$, $\quad t_0 = 1$, $\quad y(1) = 1$, $y'(1) = 2$.

4. $(y'')y' = y, t_0 = 1$ $\quad y(1) = 0$, $y'(1) = -1$.

In 5-6, find the first 5 non-zero terms of a Taylor series solution centered at zero to the damped pendulum IVP problems below.

5. $y'' + y' + \sin y = 0, y(0) = \dfrac{\pi}{2}$, $\quad y'(0) = -1$.

6. $y'' + y' + \sin y = \sin t, y(0) = \dfrac{\pi}{2}$, $y'(0) = 1$.

Chapter 7

Fourier Series and Periodic Functions

7.1 Basic Definitions and Theorems

In this chapter, we will explore Fourier Series representations for periodic functions. We will not carefully handle matters of convergence of these series as this is a basic introduction of the topic, but the reader should realize that these series may not converge.

Fourier Series

A Fourier series is a series of the form

$$\frac{a_0}{2} + \sum_{j=1}^{\infty} a_j \cos(\frac{\pi}{L} j x) + b_j \sin(\frac{\pi}{L} j x),$$

where L, a_j, and b_j are constants.

A Fourier series is similar to a Taylor series in that it is a 'function' of x. It is quite clear that if a Fourier series converges at x then it will have the same value at $x + 2L$ since sine and cosine are $2\pi-$periodic. Therefore, if the Fourier series converges for all x in $[-L, L]$ then it will converge for all x to a periodic function with period $2L$.

Next, given a periodic function, can we obtain a Fourier series that represents this function? The next result gives formulas to for the coefficients of a particular periodic function. Again, note that the Fourier series that we obtain may not converge to the original function.

Fourier Coefficients for a Given Periodic Function

Suppose that $f(x)$ is a periodic function with period $2L > 0$. That is $f(x+2L) = f(x)$ for all x. Then

$$a_0 = \frac{1}{L} \int_{-L}^{L} f(x) \, dx$$

$$a_j = \frac{1}{L} \int_{-L}^{L} f(x) \cos(\frac{\pi}{L} j x) \, dx$$

$$b_j = \frac{1}{L} \int_{-L}^{L} f(x) \sin(\frac{\pi}{L} j x) \, dx$$

Moreover, if $f(x)$ is continuous on $[-L, L]$ and has continuous derivative on $(-L, L)$ then the Fourier series converges uniformly to $f(x)$.

As you can see, there are an infinite number of coefficients to compute, but j is constant with respect to the integral, so this can often be done.

Example 7.1 *Verify that the Fourier coefficients of*

$$f(x) = 3 \sin(2x)$$

are $a_1 = 3$ and all the rest equal to zero, where $L = \pi$.

Solution: First,

$$a_0 = \frac{1}{\pi} \int_{-\pi}^{\pi} 3 \sin(x) \, dx = \frac{3}{\pi} - \cos(x)\big|_{-\pi}^{\pi} = 0.$$

$$a_1 = \frac{1}{\pi} \int_{-\pi}^{\pi} 3 \sin^2(x) \, dx = \frac{3}{\pi} \int_{-\pi}^{\pi} \frac{1}{2} - \frac{1}{2} \cos(2x) \, dx = \frac{3}{\pi} \left(\frac{1}{2} x - \frac{1}{4} \sin(2x) \right) \Big|_{-\pi}^{\pi}.$$

$$= \frac{3}{\pi} \frac{1}{2} (\pi - (-\pi)) - \frac{1}{4} \sin(2\pi) + \frac{1}{4} \sin(2(-\pi)) = 3.$$

To compute a_j for $j > 1$ we use the product to sum formula:

$$\sin(w) \sin(z) = \frac{1}{2} \cos(w - z) - \frac{1}{2} \cos(w + z)$$

$$a_j = \frac{1}{\pi} \int_{-\pi}^{\pi} 3\sin(x)\sin(jx) \, dx = \frac{3}{\pi} \int_{-\pi}^{\pi} \frac{1}{2}\cos[(1-j)x] - \frac{1}{2}\cos[(1+j)x] \, dx$$

$$= \frac{3}{\pi} \left(\frac{1}{2}\frac{1}{1-j}\sin[(1-j)x] - \frac{1}{2}\frac{1}{1+j}\sin[(1+j)x] \right) \Big|_{-\pi}^{\pi} = 0$$

$$b_j = \frac{1}{\pi} \int_{-\pi}^{\pi} 3\sin(x)\cos(jx) \, dx = \frac{3}{\pi} \int_{-\pi}^{\pi} \frac{1}{2}\sin[(1+j)x] + \frac{1}{2}\sin[(1-j)x] \, dx$$

$$= \frac{3}{\pi} \left(-\frac{1}{2}\frac{1}{1+j}\cos[(1+j)x] + \frac{1}{2}\frac{1}{1-j}\cos[(1-j)x] \right) \Big|_{-\pi}^{\pi} = 0.$$

The above quantity is zero since $\cos[(1 \pm j)\pi] = \cos[(1 \pm j)(-\pi)]$ since cosine is an even function.

$\square$

Similar to the previous example, one can use trigonometric formulas to compute the $2L$−periodic Fourier coefficients for any linear combination of $\cos(\frac{n\pi}{L}x)$ and $\sin(\frac{n\pi}{L}x)$, the Fourier coefficients will match the weights of the linear combination.

Fourier Coefficients for Linear Combinations of $\cos(\frac{n\pi}{L}x)$ and $\sin(\frac{n\pi}{L}x)$
Suppose that $f(x)$ is a linear combination of $\cos(\frac{n\pi}{L}x)$ and $\sin(\frac{n\pi}{L}x)$, then the Fourier coefficients of $f(x)$ with period $2L$ are exactly equal to the coefficients. That is, the original function is already expressed as a Fourier series.

The result follows from the orthonormality formulas, which are each easily proven, which we will simply state:

Fourier Series Orthonormality Formulas: For nonnegative integers j, k.

$$\frac{1}{L} \int_{-L}^{L} \cos(\frac{\pi}{L}jx) \cos(\frac{\pi}{L}jx) \, dx = 1$$

$$\frac{1}{L} \int_{-L}^{L} \sin(\frac{\pi}{L}jx) \sin(\frac{\pi}{L}jx) \, dx = 1$$

$$\frac{1}{L} \int_{-L}^{L} \sin(\frac{\pi}{L}kx) \sin(\frac{\pi}{L}jx) \, dx = 0, j \neq k$$

$$\frac{1}{L} \int_{-L}^{L} \cos(\frac{\pi}{L}kx) \cos(\frac{\pi}{L}jx) \, dx = 0, j \neq k$$

$$\frac{1}{L} \int_{-L}^{L} \cos(\frac{\pi}{L}kx) \sin(\frac{\pi}{L}jx) \, dx = 0$$

Example 7.2 *Consider $f(x) = x$ on $(-1, 1]$ and repeated to be periodic with period 2 (see Figure). Compute the Fourier coefficients and plot the finite fourier series with four terms and with thirty terms.*

Solution: First note that $L = 1$.

So

$$a_0 = \int_{-1}^{1} x \, dx = \frac{x^2}{2} \Big|_{-1}^{1} = 0.$$

Next, for $j \geq 1$ we use integration by parts to obtain

$$a_j = \int_{-1}^{1} x \cos(j\pi x) \, dx = \frac{1}{\pi j} x \sin(j\pi x) \Big|_{-1}^{1} + \frac{1}{\pi^2 j^2} \cos(j\pi x) \Big|_{-1}^{1} = 0$$

(the first term is zero since j is an integer and the second since cosine is even).

$$b_j = \int_{-1}^{1} x \sin(j\pi x) \, dx = -\frac{1}{\pi j} x \cos(j\pi x) \Big|_{-1}^{1} + \frac{1}{\pi^2 j^2} \sin(j\pi x) \Big|_{-1}^{1} = -\frac{1}{\pi j} 2 \cos(j\pi)$$

$$= (-1)^{j+1} \frac{1}{\pi j}.$$

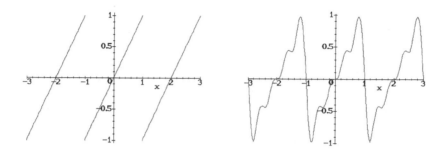

Figure 7.1: $f(x)$ and its four term Fourier Series approximation

We plot four terms of the Fourier series of $f(x)$, namely

$$\frac{2}{\pi}\sin(\pi x) - \frac{2}{2\pi}\sin(2\pi x) + \frac{2}{3\pi}\sin(3\pi x) - \frac{2}{4\pi}\sin(4\pi x).$$

We also plot the 30-term Fourier series approximation.

$\square$

In the previous example, we see that the Fourier series converges to the function except at the endpoint $x = 1$ (the original function had value of 1 at $x = 1$ and the extended periodic function is discontinuous at $x = 1$ and $x = -1$). It turns out that the Fourier series will converge at a discontinuity c of $f(x)$ to the average of the left and right hand limits of the function at $x = c$, namely:

Fourier Series Convergence

Suppose that $f(x)$ a piecewise continuous periodic function with period $2L$, then if $f(x)$ is continuous at x the Fourier series converges to $f(x)$. If $f(x)$ is discontinuous at $x = c$ then the Fourier series will converge to

$$\frac{1}{2}\left(\lim_{x \to c^-} f(x) + \lim_{x \to c^+} f(x)\right)$$

In the previous example, we see that all of the a_j terms are zero. This is unsurprising, since the function $f(x)$ was odd on $(-1, 1)$. That is $f(-x) = -f(x)$ for all x in $(-1, 1)$.

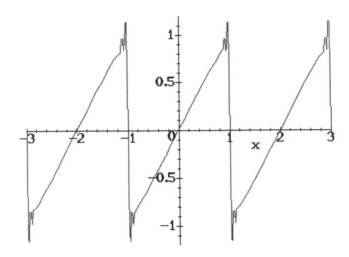

Figure 7.2: The thirty term Fourier Series approximation

<u>Issues That Arise Using Technology to Integrate</u>

Clearly, If $f(x) = \sin(\pi x)$ on $(-1, 1)$ then we expect that when we perform the integrals, to get $b_1 = 1$ and all others are equal to zero.

But computing $b_j = \int_{-1}^{1} \sin(\pi x) \sin(j\pi x) \, dx$ on a computer algebra system this integral gives:

$$b_j = -\frac{2\sin(j\pi)}{\pi(j^2 - 1)}.$$

This is UNDEFINED for $j = 1$ (and zero for all integers $j \geq 2$). The computer ignored the case $j = 1$ and performed the computation for all real constants j. To get the correct answer when $j = 1$ we can take a limit and use L'Hopital's rule to recover the correct answer:

$$b_1 = \lim_{j \to 1} -\frac{2\sin(j\pi)}{\pi(j^2 - 1)} = \lim_{j \to 1} -\frac{2\cos(j\pi)\pi}{2j\pi} = 1$$

Exercises

In 1-2, find the Fourier coefficients for the following functions which are extended to be periodic from the interval on which they are defined.

1. $f(x) = 4\sin(5x), -\pi < x \leq \pi$

2. $f(x) = \cos x - 3\sin(3x), -\pi < x \leq \pi$

In 3-6, find the Fourier coefficients for the following functions which are extended to be periodic from the interval on which they are defined. Use technology to graph the periodic function and the finite Fourier series approximations for $n = 4, 6, 8$ showing at least two periods of the function

3. $f(x) = \sin x, -\frac{\pi}{2} < x \leq \frac{\pi}{2}$

4. $f(x) = e^x, -\pi < x \leq \pi$

5. $f(x) = |x|, -\pi < x \leq \pi$

6. $f(x) = \frac{|x|}{x}, -\pi < x \leq \pi$

7.2   Fourier Series for Even and Odd Functions

Recall,

Even and Odd Functions

Suppose that $f(-x) = f(x)$ for all x then f is called an even function.
Suppose that $f(-x) = -f(x)$ for all x then f is called an odd function.

The following facts are easily checked:

1. the product of two even functions is even.

2. the product of two odd functions is even.

3. the product of an even and an odd function is odd.

The above facts can be used to show:

Fourier Series for Even Function

Suppose that $f(-x) = f(x)$ for all x and suppose that $f(x)$ is periodic with period $2L$.
Then for all j :

$$b_j = 0,$$

$$a_j = \frac{2}{L} \int_0^L f(x) \cos(\frac{\pi}{L} jx)\ dx$$

and

$$a_0 = \frac{2}{L} \int_0^L f(x)\ dx$$

Fourier Series for Odd Function

Suppose that $f(-x) = -f(x)$ for all x and suppose that $f(x)$ is periodic with period $2L$.
Then for all j :

$$a_j = 0,$$

$$b_j = \frac{2}{L} \int_0^L f(x) \sin(\frac{\pi}{L} jx)\ dx$$

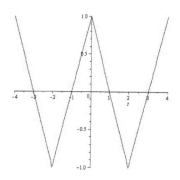

Figure 7.3: $f(x)$

Example 7.3 *Consider* $f(x) = x - 1$ *on* $(0, 2)$. *Extend this to be an even function on* $(-2, 2)$ *and then extend this to be a periodic function with period 4 on the entire real line.*

  (a) Plot this function on $[-4, 4]$

  (b) Compute the Fourier coefficients.

  (c) Plot the finite Fourier series approximations for $n = 4$ *and* $n = 100$

Solution: (a) See figure.

  (b) To compute the Fourier coefficients, we note that this function is even, so all $b_j = 0$.

$$a_0 = \frac{2}{2} \int_0^2 x - 1 \; dx = \left. \frac{x^2}{2} - x \right|_0^2 = 2 - 2 = 0$$

$$a_j = \frac{2}{2} \int_0^2 \cos(\frac{\pi}{2} j x)(x - 1) \; dx$$

after integrating by parts, we obtain:

$$a_j = \frac{4}{\pi^2 j^2} \left(1 - (-1)^j \right)$$

  (c) See figure.

$\square$

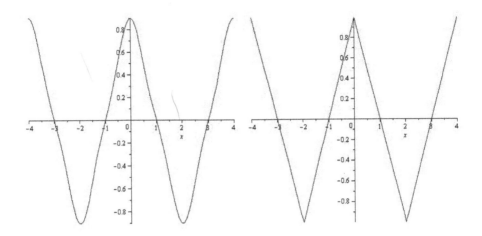

Figure 7.4: The four-term and 100-term Fourier approximations

Example 7.4 *Consider* $f(x) = x - 1$ *on* $(0, 2)$. *Extend this to be an odd function on* $(-2, 2)$ *and then extend this to be a periodic function with period 4 on the entire real line.*

    (a) Plot this function on $[-4, 4]$

    (b) Compute the Fourier coefficients.

    (c) Plot the finite Fourier series approximations for $n = 4$ *and* $n = 100$

Solution: (a) See figure.

    (b) To compute the Fourier coefficients, we note that this function is odd, so all $a_j = 0$.

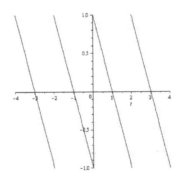

Figure 7.5: $f(x)$

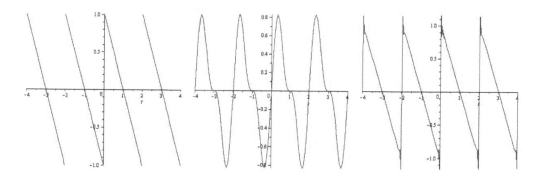

Figure 7.6: $f(x)$, the four-term, and 100-term Fourier approximations

$$b_j = \frac{2}{2} \int_0^2 \sin(\frac{\pi}{2}jx)(x-1) \ dx$$

after integrating by parts, we obtain:

$$b_j = \frac{2}{\pi j}\left(1 + (-1)^j\right)$$

(c) See figure.

$\square$

Exercises

In 1-6, find the Fourier coefficients for the following functions which are extended to be periodic from the interval on which they are defined. Use technology to graph the original function and the finite Fourier series approximations for $n = 4, 20$ showing at least two periods of the function

1. $f(x) = \sin x, 0 < x \leq \frac{\pi}{2}$ extend to a π- periodic even function.

2. $f(x) = e^x, 0 < x \leq \pi$ extend to a 2π- periodic odd function.

3. $f(x) = x, 0 < x \leq 5$ extend to a period 10 even function.

4. $f(x) = 1, 0 < x \leq \pi$ extend to a period 2π odd function.

5. $f(x) = 1, 0 < x \leq \pi$ extend to a period 2π even function.

6. $f(x) = x^2, 0 < x \leq \pi$ extend to a period 2π odd function.

7.3 Root Mean Square (RMS), Energy, Resonance

In this section, we will make use of the following identities that are easily verified:

$$\frac{1}{L}\int_{-L}^{L}\sin(\frac{\pi}{L}kx)\sin(\frac{\pi}{L}jx)\ dx = \begin{cases} 0 & j \neq k \\ \\ 1 & j = k \end{cases}$$

$$\frac{1}{L}\int_{-L}^{L}\cos(\frac{\pi}{L}kx)\cos(\frac{\pi}{L}jx)\ dx = \begin{cases} 0 & j \neq k \\ \\ 1 & j = k \end{cases} \tag{7.1}$$

$$\frac{1}{L}\int_{-L}^{L}\sin(\frac{\pi}{L}kx)\cos(\frac{\pi}{L}jx)\ dx = 0$$

The root means square for a function $f(x)$ on the interval $[-L, L]$ is

$$RMS = \sqrt{\frac{1}{2L}\int_{-L}^{L}[f(x)]^2\ dx}.$$

When a function on $[-L, L]$ is extended to be periodic, and is expressed as a Fourier Series, the RMS is very easy to compute.

Since

$$RMS = \frac{1}{\sqrt{2}}\sqrt{\frac{1}{L}\int_{-L}^{L}\left(\frac{a_0}{2} + \sum_{j=1}^{\infty}a_j\cos(\frac{\pi}{L}jx) + b_j\sin(\frac{\pi}{L}jx)\right)^2\ dx}$$

so

$$2(RMS)^2 =$$

$$\frac{1}{L}\int_{-L}^{L}\left(\frac{a_0^2}{4} + \sum_{j=1}^{\infty}\left(a_j^2\cos^2(\frac{\pi}{L}jx) + (b_j)^2\sin^2(\frac{\pi}{L}jx)\right) + \sum_{j\neq k}a_jb_k\cos(\frac{\pi}{L}jx)\sin(\frac{\pi}{L}kx)\right)\ dx$$

By the above results, this simplifies to

$$2(RMS)^2 = \frac{a_0^2}{2} + \sum_{j=1}^{\infty}a_j^2 + b_j^2$$

Example 7.5 *(a) Express* $(RMS)^2$ *for the function on* $[-2, 2]$

$$f(x) = \begin{cases} 1 & 0 < x < 2 \\ -1 & -2 < x < 0 \end{cases}$$

in terms of a series.

(b) Explain why the function in (a) has the same $(RMS)^2$ *as*

$$g(x) = 1$$

on $[-2, 2]$, *and show that the* $(RMS)^2$ *of* $g(x) = 1$ *on* $[-2, 2]$ *is 1.*

(c) Use (a) and (b) to find $\sum_{j=1}^{\infty} \frac{1}{(2j-1)^2}$.

Solution:

(a) We express $f(x)$ as a Fourier series. It is clearly an odd function, so $a_n = 0$.

$$b_n = \frac{2}{2} \int_0^2 \sin(\frac{n\pi}{2}t)dt$$

$$= -\cos(\frac{n\pi}{2}t)\frac{2}{n\pi}\Big|_{t=0}^{2}$$

$$= -\cos(n\pi)\frac{2}{n\pi} + \frac{2}{n\pi}$$

$$= \frac{2}{n\pi}(1 - (-1)^n).$$

Which is zero for n even and $\frac{4}{n\pi}$ for n odd. So $b_{2j-1} = \frac{4}{\pi(2j-1)}$.

$$2(RMS)^2 = \sum_{k=1}^{\infty} b_k^2 = \sum_{j=1}^{\infty} \frac{16}{\pi^2(2j-1)^2}.$$

$$(RMS)^2 = \frac{8}{\pi^2} \sum_{j=1}^{\infty} \frac{1}{(2j-1)^2}.$$

(b) Since $(f(x))^2 = (g(x))^2$ they have the same value for RMS and $(RMS)^2$. Moreover, RMS of $g(x)$ is clearly equal to 1.

(c) From

$$(RMS)^2 = \frac{8}{\pi^2} \sum_{j=1}^{\infty} \frac{1}{(2j-1)^2}$$

we obtain

$$1 = \frac{8}{\pi^2} \sum_{j=1}^{\infty} \frac{1}{(2j-1)^2}$$

so

$$\sum_{j=1}^{\infty} \frac{1}{(2j-1)^2} = \frac{\pi^2}{8}$$

$\square$

Resonance in a Spring Mass System–Revisited

Recall that in a spring mass system, the energy is given by:

$$E(t) = \frac{1}{2}m\left[y'(t)\right]^2 + \frac{1}{2}k\left[y(t)\right]^2.$$

We also saw earlier that if one uses functions of the form $F_{ext}(t) = F_0 \cos(\omega t - \phi)$, then the optimal forcing frequency in an underdamped system is $\omega = \frac{\sqrt{2mk-b^2}}{\sqrt{2m}}$. In this section, we will show that this form of function yields the optimal energy response for the steady state solution over *all* possible periodic forcing functions.

Assume that $F_{ext}(t)$ is a periodic function with period $2T$ having a fixed amount of average energy of E_* on $[-T, T]$, that is we assume that

$$E_* = k\frac{1}{T} \int_{-T}^{T} [F_{ext}(t)]^2 \; dt.$$

We extend $F_{ext}(t)$ as a Fourier series, which expresses as:

$$F_{ext}(t) = \frac{e_0}{2} + \sum_{j=1}^{\infty} e_n \cos(\frac{n\pi}{T}t) + f_n \sin(\frac{n\pi}{T}t).$$

Thus, the average energy can be computed using Formulas (7.1), and is: $E_* = k\frac{e_0^2}{2} + k\sum_{n=1}^{\infty}(e_n^2 + f_n^2)$.

For any forcing terms of the form $e_n \cos(\omega t) + f_n \sin(\omega t)$ we can use undetermined coefficients to find the steady state solution which is of the form $A_n \cos(\omega t) + B_n \sin(\omega t)$.

It turns out that

$$A_n = -\frac{mw^2 e_n + bw f_n - k e_n}{b^2 w^2 + m^2 w^4 - 2mw^2 k + k^2}$$

$$B_n = -\frac{mw^2 f_n - k f_n - e_n bw}{b^2 w^2 + m^2 w^4 - 2mw^2 k + k^2}$$

So for

$$\omega = \frac{\pi n}{T}$$

we obtain:

$$y(t) = A_n \cos(\frac{\pi n}{T} t) + B_n \sin(\frac{\pi n}{T} t)$$

and

$$y'(t) = -A_n \left(\frac{\pi n}{T}\right) \sin(\frac{\pi n}{T} t) + B_n \left(\frac{\pi n}{T}\right) \cos(\frac{\pi n}{T} t).$$

Therefore,

$$y_{steady}(t) = \frac{e_0}{2k} + \sum_{n=1}^{\infty} A_n \cos(\frac{\pi n}{T} t) + B_n \sin(\frac{\pi n}{T} t)$$

The average energy of this solution over $[-T, T]$ is

$$\frac{1}{2T} \int_{-T}^{T} E(t)\ dt = \frac{1}{2T} \int_{-T}^{T} \frac{1}{2} m\left[y'(t)\right]^2 + \frac{1}{2} k\left[y(t)\right]^2\ dt.$$

$$= \frac{m}{4T} \int_{-T}^{T} \left[\sum_{n=1}^{\infty} -A_n \left(\frac{\pi n}{T}\right) \sin(\frac{\pi n}{T} t) + B_n \left(\frac{\pi n}{T}\right) \cos(\frac{\pi n}{T} t)\right]^2\ dt$$

$$+ \frac{k}{4T} \int_{-T}^{T} \left[\frac{e_0}{2k} + \sum_{n=1}^{\infty} A_n \cos(\frac{\pi n}{T} t) + B_n \sin(\frac{\pi n}{T} t)\right]^2\ dt.$$

It is now that we make use of the integral formulas given at the beginning of this section.

$$= \frac{m}{4} \left(\frac{\pi n}{T}\right)^2 \sum_{n=1}^{\infty} (A_n^2 + B_n^2) + \frac{k}{4} \left(\frac{e_0^2}{2k^2} + \sum_{n=1}^{\infty} (A_n^2 + B_n^2)\right),$$

which (since $A_n^2 + B_n^2$ simplifies) can be written as:

$$E = \frac{e_0^2}{8k} + \frac{1}{4}\sum_{n=1}^{\infty} \frac{T^2(f_n^2 + e_n^2)(\pi^2 n^2 m + kT^2)}{b^2 n^2 \pi^2 T^2 + m^2 n^4 \pi^4 - 2mn^2 \pi^2 kT^2 + k^2 T^4}$$

which is equivalent to maximizing:

$$4E = \frac{1}{k}\frac{e_0^2}{2} + \sum_{n=1}^{\infty} \frac{T^2(\pi^2 n^2 m + kT^2)}{b^2 n^2 \pi^2 T^2 + (\pi^2 n^2 m - kT^2)^2} e_n^2$$

$$+ \sum_{n=1}^{\infty} \frac{T^2(\pi^2 n^2 m + kT^2)}{b^2 n^2 \pi^2 T^2 + (\pi^2 n^2 m - kT^2)^2} f_n^2$$

subject to: $\frac{E_*}{k} = \frac{e_0^2}{2} + \sum_{n=1}^{\infty} e_n^2 + \sum_{n=1}^{\infty} f_n^2$.
When rewrite

$$4E = x_0 \frac{e_0^2}{2} + \sum_{n=1}^{\infty} x_n e_n^2 + \sum_{n=1}^{\infty} x_n f_n^2$$

where $x_0 = \frac{1}{k}$ and

$$x_n = \frac{T^2(\pi^2 n^2 m + kT^2)}{b^2 n^2 \pi^2 T^2 + (\pi^2 n^2 m - kT^2)^2}$$

it is clear that $4E$ is maximized by choosing the maximum of the x_n and assigning all other e_j and $f_j = 0$. Note that $x_n > 0$ and $x_n \to 0$ as $n \to \infty$ so there is a maximum value for x_n. Let N be one such maximal value.

Thus, a solution producing optimal average steady state energy is either a constant function or of the form

$$y(t) = A_N \cos(\frac{\pi N}{T} t) + B_N \sin(\frac{\pi N}{T} t).$$

However, we already explored these types of functions earlier and found the optimal frequency to be $\omega = \frac{\sqrt{2mk - b^2}}{\sqrt{2m}}$ for lightly underdamped spring mass systems (which is when $b^2 - 2mk < 0$), otherwise constant forcing produces maximal average energy (which is when $b^2 - 2mk \geq 0$).

We summarize our results:

Resonance in a Forced Spring/Mass Systems

The periodically forced spring mass system

$$my'' + by' + ky = F_{external}(t) \tag{7.2}$$

exhibits resonance when $b^2 - 2mk < 0$ (we call such a system lightly damped). In such cases, the steady state solution with optimal output energy is obtained by external forcing functions (having fixed input energy) of the form: $F_{external}(t) = F_0 \cos(\omega t - \phi)$ where

$$\omega = \frac{\sqrt{2mk - b^2}}{\sqrt{2m}}. \tag{7.3}$$

If $b^2 - 2mk \geq 0$ then the steady state solution with optimal output energy is obtained by a constant function.

The sequence x_n derived above can itself be useful as the following example demonstrates.

Example 7.6 *Consider a spring mass system with $m = 1$, $b = 1$ and $k = 2$. Find the optimal periodic external forcing function that has fixed energy, has period 24, and produces the optimal steady state solution with respect to energy.*

Solution: We plot the sequence x_n derived above (note $T = 12$, not 24) and find that x_5 is the maximum value of this sequence (see figure). Thus, optimal forcing functions (for period 24) are of the form $\cos(\frac{5\pi}{12}t - \phi)$.

$\square$

Note that in the previous example, $\frac{5\pi}{12}$ is the closest possible frequency of the form $\frac{n\pi}{12}$ to the true resonance frequency $\frac{\sqrt{2mk-b^2}}{\sqrt{2m}}$.

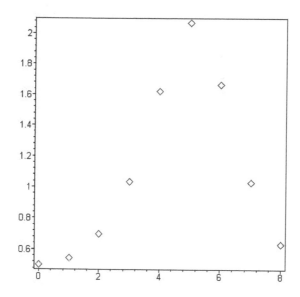

Figure 7.7: The sequence x_n for $m = 1$, $b = 1$, $T = 12$, and $k = 2$.

Exercises

1. Compute the root mean square of a constant function $f(t) = a$.

2. Compute the root mean square of $f(t) = a\sin(\beta t)$ (period is $\frac{2\pi}{\beta}$).

3. Compute the RMS of a square wave with amplitude $a > 0$ and period $2L$
   namely $f(t) = \begin{cases} -a & -L < t < 0 \\ a & 0 < t < L \end{cases}$

4. Compute the root mean square of $f(t) = t$ for $0 < t < 1$ extended to be
   an even function with period 2.

5. Plot the sequence x_n and use it to find the periodic external forcing function
   with fixed energy of $E_* = 1$ that has steady state solution with maximal
   energy, has period 10, for a spring mass system with $m = 1$, $b = 0.5$, and
   $k = 6$. Find the optimal amplitude for such a steady state.

6. Plot the sequence x_n and use it to find the periodic external forcing function
   with fixed energy of $E_* = 1$ that has steady state solution with maximal

energy, has period 7, for a spring mass system with $m = 1$, $b = 2$, and $k = 3$. Find the optimal amplitude for such a steady state.

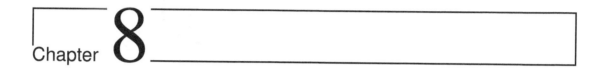

Chapter 8

Partial Differential Equations

8.1   The Heat Equation

In this section, we consider a temperature of a bar at time t. We assume the bar is of length $L > 0$. Imagine the bar is placed on the number line from $[0, L]$ we wish to consider how heat is diffused throughout the bar.

Assume

(1) The temperature on any cross-section of the bar is the same on the entire cross-section. That is, temperature only depends on 2 things: time and location on the x-axis.

(2) Heat is transferred only within the bar itself, and may only escape (enter) the bar through its ends.

In light of these assumptions, The temperature of the bar depends on two things: 1. what time it is; and 2. where on the bar we wish to consider. In other words, we consider a function of two variables:

$$u(x, t) = \text{temperature at position } x \text{ at time } t$$

for $0 \leq x \leq L$. An initial temperature distribution at time $t = 0$ is required similar to initial conditions for an ODE $u(x, 0) = f(x)$ for $0 < x < L$. Also behavior at the ends of the bar must also be specified (e.g. is heat escaping?).

Firstly, we must understand how temperature changes with time at any point (x, t). In other words, we seek to understand $\frac{\partial u}{\partial t}$ at (x, t). Certainly, the actual time it is has no effect (heat transfers at the same rate independent of the time), so we expect $\frac{\partial u}{\partial t}$ to only depend on the temperatures near x and the material of

247

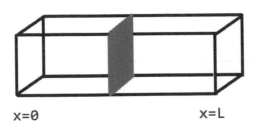

$x=0$ $x=L$

Figure 8.1: A bar of length L with cross-section. We assume the temperature on any cross-section is uniformly the same and the heat can only be transferred into/out of the bar at the ends.

makeup of the rod. Secondly, we assume that heat is transmitted (conducted) at a rate up to some constant. This constant changes as the units with which temperature and time are measured are changed and also depends on how well the bar can conduct heat relative to these units of time and temperature. So we will find an equation for $\frac{\partial u}{\partial t}$ up to a constant β. The constant β is called the constant of thermal diffusivity. A more conductive material will have a higher value of β (and hence, lower values of β amount to a faster rate of heat transfer).

Aside from this constant, the most important consideration that determines the temperature change at location x is the temperatures near x. That is, if it is hotter to the right of x and to the left, then we expect $\frac{\partial u}{\partial t} > 0$. If it is hotter to the right and colder to the left, we expect some averaging. In particular, if we average these differences (over Δx) $\frac{u(x-\Delta x,t)-u(x,t)}{\Delta x}$ and $\frac{u(x+\Delta x,t)-u(x,t)}{\Delta x}$ we obtain a good approximation for what Δu is proportional to, namely:

$$\frac{\frac{u(x-\Delta x,t)-u(x,t)}{\Delta x} + \frac{u(x+\Delta x,t)-u(x,t)}{\Delta x}}{\Delta x},$$

or

$$\Delta u \propto \frac{u(x - \Delta x, t) + u(x + \Delta x, t) - 2u(x,t)}{(\Delta x)^2}.$$

As $\Delta x \to 0$ the above term limits to the second partial $u_{xx}(x,t)$. We have heuristically derived the heat equation (a more formal derivation comes from Fourier's Law).

> ### The Heat Equation
> In a bar with length L and temperature $u(x,t)$ for $0 \le x \le L$, we have
>
> $$u_t(x,t) = \beta u_{xx}(x,t) \qquad (8.1)$$
>
> where $\beta > 0$ is a constant called the constant of thermal diffusivity.

One thing that is clear is that this DE is linear. Meaning if $u_1(x,t)$ and $u_2(x,t)$ solve this PDE, then so will $c_1 u_1(x,t) + c_2 u_2(x,t)$.

We can observe several 'easy' solutions to this PDE. Namely, $u(x,t) = 0$ (or any other constant) solves this PDE. Note that $u(x,t) = x$ is also a solution to the PDE, but $u(x,t) = t$ is not.

In particular, the steady state solutions (i.e., solutions that remain constant with respect to time) to the heat equation (where $u_t = 0$) are clearly $u(x,t) = Ax + B$.

Method of Separation of Variables

Next, we consider a method that can be used to find certain solutions to partial differential equations that are can be expressed as $X(x)T(t)$ where $X(x)$ is a function of x only and $T(t)$ is a function of t only. That is, we seek solutions of the PDE that can be written in the form

$$u(x,t) = X(x)T(t).$$

(For instance $x^2 \cos t$ is such a product where $X(x) = x^2$ and $T(t) = \cos t$.) We will demonstrate this technique on the heat equation $u_t = \beta u_{xx}$ together with conditions $u(0,t) = 0$ and $u(L,t) = 0$ (meaning we are holding the ends at a fixed temperature of 0).

If a separable solution $u(x,t) = X(x)T(t)$ exists to this PDE, then:

$$\beta u_{xx} = \beta X''(x)T(t)$$

and

$$u_t = X(x)T'(t)$$

and so (by plugging into the DE)

$$X(x)T'(t) = \beta X''(x)T(t)$$

Separating (when functions are nonzero) we obtain:

$$\frac{T'(t)}{\beta T(t)} = \frac{X''(x)}{X(x)}.$$

In other words, we have an expression on the left that only depends on t and and expression on the right that only involves x in. Since the expression on the left cannot depend on x (it only involves t), and the expression on the right cannot involve t (it only involves x), both expressions must be equal to a constant which we will label as $-\lambda$ (it will be clear later why we choose this notation).

That means we obtain two ordinary differential equations

$$\frac{T'(t)}{\beta T(t)} = -\lambda$$

and

$$\frac{X''(x)}{X(x)} = -\lambda$$

If $T(t) = 0$ or $X(x) = 0$, then we have the solution $u(x,t) = 0$. We are searching for non-zero solutions. So we realize that we obtain two ordinary differential equations:

$$T'(t) + \beta\lambda T(t) = 0$$

and

$$X''(x) + \lambda X(x) = 0.$$

Each of these differential equations are easily solved, depending on whether λ is zero, positive, or negative.

Case 1: $\lambda = 0$. In this case,

This means

$$\frac{X''(x)}{X(x)} = 0$$

implies that

$$X''(x) = 0.$$

This has general solution

$$X(x) = c_1 x + c_2$$

Since $X(0) = 0$ we see that $c_2 = 0$ and $X(L) = 0$ yields $c_1 L = 0$, which forces $c_1 = 0$.

This means the only possible separable solution with $\lambda = 0$ satisfying these boundary conditions is $X(x) = 0$ so $u(x, t) = 0$.

Case 2: $\lambda < 0$. Note that $-\lambda > 0$

This means

$$\frac{X''(x)}{X(x)} = -\lambda$$

implies that

$$X''(x) + \lambda X(x) = 0.$$

This has solutions

$$X(x) = c_1 e^{\sqrt{-\lambda}x} + c_2 e^{-\sqrt{-\lambda}x}$$

where $-\lambda > 0$.

Since $X(0) = 0$ we see that $c_1 + c_2 = 0$ and $X(L) = 0$ yields $c_1 e^{\sqrt{-\lambda}L} + c_2 e^{-\sqrt{-\lambda}L} = 0$.

Since $c_1 = -c_2$ we see that

$$-c_2 e^{\sqrt{-\lambda}L} + c_2 e^{-\sqrt{-\lambda}L} = 0$$

or

$$c_2(e^{-\sqrt{-\lambda}L} - e^{\sqrt{-\lambda}L}) = 0$$

The term in the parentheses is zero only when $-\lambda = 0$ or $L = 0$ (but $R > 0$ and $L > 0$ by assumption) so we must have $c_1 = c_2 = 0$, so this means the only possible separable solution with $\lambda < 0$ satisfying these boundary conditions is $X(x) = 0$ so $u(x, t) = 0$.

Case 3: $\lambda > 0$. In this case,

This means

$$\frac{X''(x)}{X(x)} = -\lambda$$

implies that

$$X''(x) + \lambda X(x) = 0.$$

This has solutions

$$X(x) = c_1 \cos(\sqrt{\lambda}x) + c_2 \sin(\sqrt{\lambda}x)$$

Since $X(0) = 0$ we see that $c_1 = 0$ and $X(L) = 0$ yields $0 = X(L) = c_2 \sin(\sqrt{\lambda}L)$.

This has solution $c_2 = 0$ (which again lead to $u(x,t) = 0$) or whenever $\sqrt{\lambda}L = k\pi$ where k is a positive integer (the zeros of sine are $k\pi$ for integers k). Therefore when

$$\lambda = \frac{k^2\pi^2}{L^2},$$

we have a solution which satisfies $X(0) = 0$ and $X(L) = 0$

This solution is

$$X(x) = c_2 \sin(\frac{k\pi}{L}x)$$

The associated solution to

$$\frac{T'(t)}{\beta T(t)} = -\lambda = -\frac{k^2\pi^2}{L^2}$$

is $T(t) = Ke^{-\frac{\beta k^2\pi^2}{L^2}t}$.

Thus there are infinitely many separable solutions of the form

$$u_k(x,t) = C_k e^{-\frac{\beta k^2\pi^2}{L^2}t} \sin(\frac{k\pi}{L}x)$$

for $k = 1, 2, 3, \ldots$ together with the trivial solution $u_0(x,t) = 0$.

We summarize what we have derived below:

Separable Solutions to Heat Equation with Homogeneous Ends

The heat equation

$$u_t = \beta u_{xx} \tag{8.2}$$

with conditions $u(0,t) = 0$ and $u(L,t) = 0$ has separable solutions of the form:

$$u_k(x,t) = c_k e^{-\frac{\beta k^2\pi^2}{L^2}t} \sin(\frac{k\pi}{L}x)$$

for $k = 1, 2, 3, \ldots$

It is clear that the limiting behavior of all separable solutions are a constant temperature of zero on the entire bar, which makes sense.

Linear combinations of solutions in the previous example will also solve the heat equation. Moreover, since $u_k(0,t) = 0$ and $u_k(L,t) = 0$, any sum of these solutions will also satisfy the homogeneous boundary conditions. This suggests the following statement:

Heat Equation with Constant Homogeneous Ends

The heat equation

$$u_t = \beta u_{xx} \tag{8.3}$$

with conditions $u(0, t) = 0$ and $u(L, t) = 0$ has solutions of the form:

$$u(x, t) = \sum_{k=1}^{\infty} c_k e^{-\frac{\beta k^2 \pi^2}{L^2} t} \sin(\frac{k\pi}{L} x),$$

moreover, the initial temperature distribution is

$$u(x, 0) = \sum_{k=1}^{\infty} c_k \sin(\frac{k\pi}{L} x)$$

where then the constants c_k are given by the Fourier coefficients of $u(x, 0)$ expanded as an ODD function on $[-L, L]$.

If the initial temperature distribution on the bar is already in the form of an odd Fourier series, then identifying the constants is easy, otherwise the function $u(x, 0)$ must be expanded as an odd Fourier series.

Example 8.1 *Consider a bar of length 3 meters. The ends are held at a constant temperature of 0°C. Assume that $\beta = \frac{1}{2}$.*
(a) Find the solution to the heat equation with initial distribution $u(x, 0) = 2\sin(\frac{5\pi}{3} x) - \sin(\frac{7\pi}{3} x)$.
(b) Find the solution to the heat equation with initial distribution $u(x, 0) = x(x - 3)$. Plot solutions for $t = 1, 2, 3$.

Solution:
(a) Note that $f(x) = 2\sin(\frac{5\pi}{3} x) - \sin(\frac{7\pi}{3} x)$ is already in the form of an odd Fourier series expansion on $[0, 3]$ so $c_5 = 2$ and $c_7 = -1$ with all other $c_k = 0$.
So the solution is

$$u(x, t) = 2e^{-\frac{25\pi^2}{18} t} \sin(\frac{5\pi}{3} x) - e^{-\frac{49\pi^2}{18} t} \sin(\frac{7\pi}{3} x).$$

(b) To solve the problem with initial temperature distribution $u(x, 0) = f(x) = x(x - 3)$ means that we need to expand this function as an odd Fourier series on $[0, 3]$.

That is, we compute $c_k = \frac{2}{3} \int_0^3 (x^2 - 3x) \sin(\frac{\pi}{3} kx) \, dx$

After integrating by parts and simplifying, we obtain:

$$c_k = 18 \frac{-2 + \pi k \sin(\pi k) + 2\cos(\pi k)}{\pi^3 k^3}$$

Since k is an integer, $\cos(\pi k) = (-1)^k$ and $\sin(\pi k) = 0$ so

$$c_k = \frac{36}{\pi^3 k^3}[(-1)^k - 1]$$

Note that $(-1)^k - 1 = 0$ if k is even and -2 if k is odd so:

$$c_k = \begin{cases} -\frac{72}{\pi^3 k^3} & k \ \ odd \\ \\ 0 & k \ \ even \end{cases}$$

Therefore,

$$u(x,t) = \frac{-72}{\pi^3} e^{-\frac{\pi^2}{18}t} \sin(\frac{\pi}{3}x) - \frac{72}{3^3 \pi^3} e^{-\frac{3^2 \pi^2}{18}t} \sin(\frac{3\pi}{3}x) \frac{72}{5^3 \pi^3} e^{-\frac{5^2 \pi^2}{18}t} \sin(\frac{5\pi}{3}x) - \dots$$

In the Figure, we plot ten terms of the Fourier series of $u(x,t)$ for various values of t. If one had enough values, one could envision the heat distribution over time as a movie.

$\square$

Nonhomogeneous, Constant End Case

Next, we consider a bar of length L where the ends are held at constant temperatures T_1 and T_2 that are different than zero. As usual, the key is in utilizing the homogeneous solution. In other words, we wish to solve the PDE with boundary conditions

$$u_t = \beta u_{xx} \ with \ u(0,t) = T_1, \ \ u(L,t) = T_2$$

We have already seen that the steady state solution to the heat equation is

$$u_{ss}(x,t) = Ax + B.$$

The condition $u(0,t) = T_1$ means we need $u = T_1$ when $x = 0$, which implies that $B = T_1$. The condition $u(L,t) = T_2$ means we need $u = T_2$ when $x = L$. This means

$$T_2 = AL + T_1$$

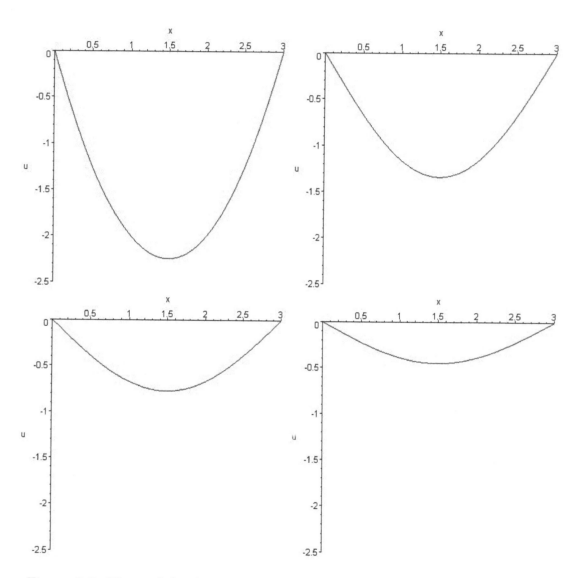

Figure 8.2: Plots of the first ten terms of $u(x,0)$,$u(x,1)$,$u(x,2)$,and $u(x,3)$.

so

$$A = \frac{T_2 - T_1}{L}.$$

So the appropriate steady state is

$$u_{ss}(x,t) = \left(\frac{T_2 - T_1}{L}\right) x + T_1.$$

We use this steady state to construct a new function $v(x,t) = u(x,t) - u_{ss}(x,t) = u(x,t) - \left(\frac{T_2 - T_1}{L}\right) x - T_1$. This new function satisfies

$$v_{xx} = u_{xx}$$

and

$$v_t = u_t,$$

so it also satisfies $v_t = \beta v_{xx}$. Moreover, the boundary conditions are $v(0,t) = 0$ and $v(L,t) = 0$. So we can solve for $v(x,t)$ as before since is a solution the the heat equation with homogeneous ends.

The only trick is that given an initial temperature distribution $u(x,0) = f(x)$ we need to translate this to $g(x) = v(x,0) = f(x) - \left(\frac{T_2 - T_1}{L}\right) x - T_1$ and we then need the Fourier coefficients of $g(x)$ to obtain the correct constants for the particular $v(x,t)$ desired. Finally, we use the fact that

$$u(x,t) = v(x,t) + u_{ss}(x,t).$$

We consolidate our findings below:

Heat Equation with Constant Nonhomogeneous Ends

The heat equation

$$u_t = \beta u_{xx} \tag{8.4}$$

with conditions $u(0,t) = T_1$ and $u(L,t) = T_2$ has solutions

$$u(x,t) = \sum_{k=1}^{\infty} c_k e^{-\frac{\beta k^2 \pi^2}{L^2} t} \sin(\frac{k\pi}{L} x) + \left(\frac{T_2 - T_1}{L}\right) x + T_1,$$

moreover, if the initial temperature distribution is $u(x,0) = f(x)$, then the constants c_k are the Fourier coefficients of

$$g(x) = f(x) - \left(\frac{T_2 - T_1}{L}\right) x - T_1$$

expanded as an ODD function on $[0, L]$.

Example 8.2 *Consider a bar of length 3 meters. The ends are held at a constant temperature of 2°C and 12°C. Assume that $\beta = 1$ and that the initial temperature distribution is $f(x) = 5°C$. Find $u(x,t)$ and plot approximations for $t = 0, 0.01, 0.1$, and 1.*

Solution:

We position the bar so that $u(0,t) = 2$ and $u(3,t) = 12$, so $u_{ss}(x,t) = \left(\frac{T_2 - T_1}{L}\right) x + T_1 = \frac{10}{3} x + 2$.

We compute the Fourier Sine coefficients of $g(x) = 5 - \frac{10}{3} x - 2 = 3 - \frac{10}{9} x$.

They are:

$$c_k = \frac{4}{k\pi}[1 + 4(-1)^k)]$$

(We used PARTS and simplified $\sin(\pi k) = 0$ and $\cos(\pi k) = (-1)^k$.)

So

$$u(x,t) = \sum_{k=1}^{\infty} \frac{4}{k\pi}[1 + 4(-1)^k)]e^{-\frac{k^2 \pi^2}{9} t} \sin(\frac{k\pi}{3} x) + \frac{10}{3} x + 2.$$

Plots are shown in the Figure.

$\square$

The previous example illustrates that the initial temperature distribution need not be continuous. Namely, we insisted that $u(x,0) = 5$ but that the ends be held at 2 and 12 respectively. Which is discontinuity at the ends when $t = 0$.

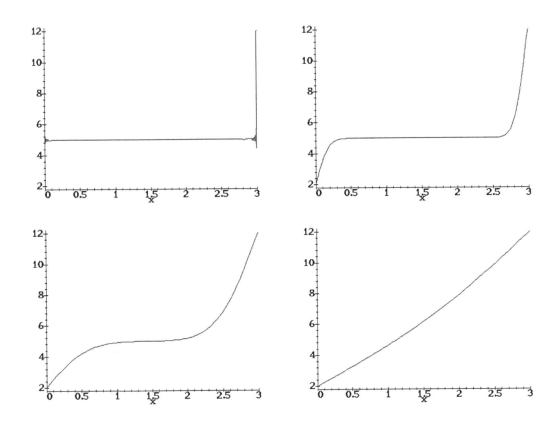

Figure 8.3: Plots of the first hundred terms of $u(x, 0)$, $u(x, 0.01)$, $u(x, 0.1)$, and $u(x, 1)$.

This allows us to specify initial temperature distributions that do not agree with the conditions placed at the ends and still obtain meaningful results.

Heat Equation with Insulated Ends

Next, we investigate the heat equation where no heat loss is allowed through the ends of a bar with length L. This means that as we approach the ends, the temperature change should limit to zero. In other words, we want solutions with $u_x(L,t) = u_x(0,t) = 0$. We will once again have to start with the separation of variables technique once again and in particular, run through the three cases that arise.

If a separable solution $u(x,t) = X(x)T(t)$ exists solving

$$u_t = \beta u_{xx}$$

then:

$$u_{xx} = X''(x)T(t)$$

and

$$\frac{1}{\beta}u_t = \frac{1}{\beta}X(x)T'(t)$$

and so

$$\frac{1}{\beta}X(x)T'(t) = X''(x)T(t)$$

Separating (when terms functions are nonzero) we obtain:

$$\beta\frac{T'(t)}{T(t)} = \frac{X''(x)}{X(x)}.$$

Since the left side of the equation only involves t and the right hand side only involves x, the both terms must be constant. So set

$$\beta\frac{T'(t)}{T(t)} = \frac{X''(x)}{X(x)} = -\lambda.$$

As usual focus on the possible solutions to

$$\frac{X''(x)}{X(x)} = -\lambda.$$

Case 1: $\lambda < 0$. Note that $-\lambda > 0$.

This means

$$\frac{X''(x)}{X(x)} = -\lambda$$

implies that

$$X''(x) + \lambda X(x) = 0.$$

This has solutions

$$X(x) = c_1 e^{\sqrt{-\lambda}x} + c_2 e^{-\sqrt{-\lambda}x}$$

where $-\lambda > 0$.

Since $u_x(0,t) = 0$ we see must have that $X'(0) = 0$, so $c_1\sqrt{-\lambda} - c_2\sqrt{-\lambda} = 0$ so $c_1 = c_2$.

Since $u_x(L,t) = 0$ we see must have that $X'(L) = 0$, so $c_1\sqrt{-\lambda}e^{\sqrt{-\lambda}L} - c_2\sqrt{-\lambda}e^{-\sqrt{-\lambda}L} = 0$.

This factors (since $c_1 = c_2$) as $c_1\sqrt{-\lambda}\left[e^{\sqrt{-\lambda}L} - e^{-\sqrt{-\lambda}L}\right] = 0$. The quantity in brackets is not zero unless $\sqrt{-\lambda}L = 0$ which is not possible in this case. So this can only be zero if $c_1 = c_2 = 0$. So, the only possible separable solution with $\lambda < 0$ is $X(x) = 0$ so $u(x,t) = 0$.

Case 2: $\lambda = 0$. In this case,

This means

$$\frac{X''(x)}{X(x)} = 0$$

implies that

$$X''(x) = 0.$$

This has solutions

$$X(x) = c_1 + c_2 x$$

Since $X'(0) = 0$ we see that $c_2 = 0$ which also satisfies $X'(L) = 0$

This means the only possible separable solution with $\lambda = 0$ is $X(x) = c_1$. Note that this implies that $T'(t) = 0$ also (since $\lambda = 0$), which means $T(t)$ is also constant. So we see that the solutions obtained from this case are $u(x,t) = C$.

Case 3: $\lambda > 0$. In this case,

This means

$$\frac{X''(x)}{X(x)} = -\lambda$$

implies that

$$X''(x) + \lambda X(x) = 0.$$

Which has solutions

$$X(x) = c_1 \cos(\sqrt{\lambda} x) + c_2 \sin(\sqrt{\lambda} x)$$

Since $X'(0) = 0$ we see that $c_2 = 0$. Since $X'(L) = 0$ we must have $X'(L) = -c_1 \sin(\sqrt{\lambda} L) = 0$ Which amounts to having $\sqrt{\lambda} L$ equalling a multiple of π. So $\sqrt{\lambda} L = k\pi$ for k a positive integer. Therefore when

$$\lambda = \frac{k^2 \pi^2}{L^2}.$$

This solution is

$$X(x) = c_1 \cos(\frac{k\pi}{L} x)$$

The associated solution to

$$\frac{1}{\beta} \frac{T'(t)}{T(t)} = -\lambda = -\frac{k^2 \pi^2}{L^2}$$

would be $T(t) = K e^{-\frac{\beta k^2 \pi^2}{L^2} t}$.

Thus there are infinitely many separable solutions of the form

$$u_k(x, t) = C_k e^{-\frac{\beta k^2 \pi^2}{L^2} t} \cos(\frac{k\pi}{3} x)$$

for $k = 1, 2, 3, \ldots$ together with $u_0(x, t) = c_0$. $\qquad \square$

Taking the solutions found in all cases we obtain:

Separable Solutions to Heat Equation with Insulated Ends

The heat equation

$$u_t = \beta u_{xx} \tag{8.5}$$

with conditions $u_x(0, t) = 0$ and $u_x(L, t) = 0$ has separable solutions of the form:

$$u_k(x, t) = c_k e^{-\frac{\beta k^2 \pi^2}{L^2} t} \cos(\frac{k\pi}{L} x)$$

for $k = 0, 1, 2, 3, \ldots$

Again it turns out that linear combination of separable solutions will also satisfy the heat equation, together with the boundary conditions $u_x(0, t) = 0$ and $u_x(L, t) = 0$, which leads us to:

Heat Equation with Insulated Ends

The heat equation

$$u_t = \beta u_{xx} \tag{8.6}$$

with conditions $u_x(0, t) = 0$ and $u_x(L, t) = 0$ has solutions of the form:

$$u(x, t) = \sum_{k=0}^{\infty} c_k e^{-\frac{\beta k^2 \pi^2}{L^2} t} \cos\left(\frac{k\pi}{L} x\right),$$

moreover, if the initial temperature distribution is $u(x, 0) = f(x)$, then the constants c_k are the Fourier coefficients of $f(x)$ expanded as an EVEN function on $[0, L]$, with the exception of c_0 (which is the a_0 Fourier coefficient of $f(x)$ divided by 2).

Exercises

1. Find the solution to heat equation with $\beta = 4$, $L = 10$ and $u(0, t) = 0, u(10, t) = 0$ having initial temperature distribution $u(x, 0) = \sin(\frac{\pi}{10}x)$.

2. Find the solution to heat equation with $\beta = 2$, $L = 2$ and $u(0, t) = 0, u(2, t) = 0$ having initial temperature distribution $u(x, 0) = 4\sin(\frac{\pi}{2}x) - 11\sin(3\frac{\pi}{2}x)$.

3. Find the solution to heat equation with $\beta = \frac{1}{2}$, $L = 1$ and $u(0, t) = 0, u(1, t) = 0$ having initial temperature distribution $u(x, 0) = 5\sin(2\pi x) - 7\sin(3\pi x)$.

4. Find the solution to heat equation with $\beta = \frac{1}{4}$, $L = 2$ and $u(0, t) = 0, u(2, t) = 0$ having initial temperature distribution $u(x, 0) = 2\sin(8\pi x) + 1\sin(16\pi x)$.

5. Find the solution to heat equation with $\beta = 1$, $L = 2$ and $u(0, t) = 0$, $u(2, t) = 0$ having initial temperature distribution $u(x, 0) = x$. Plot Fourier approximations with at least 10 terms of $u(x, 0)$, $u(x, 1)$, $u(x, 2)$, and $u(x, 3)$.

6. Find the solution to heat equation with $\beta = \frac{1}{2}$, $L = 3$ and $u(0, t) = 0$, $u(3, t) = 0$ having initial temperature distribution $u(x, 0) = 5$. Plot Fourier approximations with at least 15 terms of $u(x, 0)$, $u(x, 1)$, $u(x, 2)$, and $u(x, 3)$.

7. Find the solution to heat equation with $\beta = \frac{1}{4}$, $L = 2$ and $u(0, t) = 0$, $u(2, t) = 0$ having initial temperature distribution $u(x, 0) = \cos x$. [Some trig. identities may be helpful]

8. Find the solution to heat equation with $\beta = 1$, $L = 2$ and $u(0, t) = 32$, $u(2, t) = 0$ having initial temperature distribution $u(x, 0) = 2\sin(\frac{\pi}{2}x) + 16x - 32$.

9. For $L = 5$, $\beta = \frac{1}{2}$, $u(0, t) = 100$, $u(5, t) = 0$ and $u(x, 0) = 30 - 8\sin(6\pi x)$

10. Find the solution to heat equation with $\beta = 1$, $L = 4$ and $u(0,t) = 50$, $u(5,t) = 60$ having initial temperature distribution $u(x,0) = x$.

11. Find the solution to heat equation with $\beta = \frac{1}{2}$, $L = 5$ and $u(0,t) = 100$, $u(5,t) = 0$ having initial temperature distribution $u(x,0) = 100 - x^2$.

12. Find the solution to heat equation with $\beta = \frac{1}{2}$, $L = 3$ and $u_x(0,t) = 0$, $u_x(3,t) = 0$ having initial temperature distribution $u(x,0) = 30 - 8\cos(6\pi x)$.

13. Find the solution to heat equation with $\beta = 1$, $L = \pi$ and $u_x(0,t) = 0$, $u_x(\pi,t) = 0$ having initial temperature distribution $u(x,0) = 4 + 3\cos(4x)$.

14. Find the solution to heat equation with $\beta = \frac{1}{2}$, $L = 10$ and $u_x(0,t) = 0$, $u_x(10,t) = 0$ having initial temperature distribution $u(x,0) = x$.

15. Find the solution to heat equation with $\beta = \frac{1}{2}$, $L = 10$ and $u_x(0,t) = 0$, $u_x(10,t) = 0$ having initial temperature distribution $u(x,0) = \sin x$. Plot Fourier approximations with at least 15 terms of $u(x,0)$, $u(x,1)$, $u(x,2)$, and $u(x,3)$.

16. Solve $u_t = \beta u_{xx}$ with the mixed boundary conditions $u_x(0,t) = 0$ and $u(L,t) = 0$ (one end insulated, and one is held constant). Start by using separation of variables to find separable solutions. Run through the three cases, then obtain a general solution by taking all linear combinations of separable solutions.

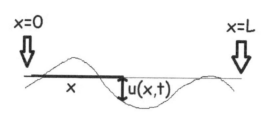

Figure 8.4: A string with imposed coordinate system from $x = 0$ to $x = L$.

8.2 The Wave Equation

The Wave Equation-Fixed Homogeneous Ends

In this section, we consider an elastic string. Imagine that a number line from $[0, L]$ is super imposed onto this picture so that $x = 0$ is associated with one end of the string and $x = L$ is associated with one end of the string as in the figure shown below.

The string is elastic so its length is not constant (it can be stretched to arclength larger than L). Unlike the heat equation, x represents the location on the number line and $u(x, t)$ is the signed (perpendicular) distance to this number line, where positive is 'upward'.

Firstly, we seek to understand the restorative force at the location and time (x, t). We label this force as $F(x, t)$. By Newton's Law $F = ma$, we seek to understand u_{tt} at (x, t) which is proportional to $F(x, t)$.

For a given $\Delta x > 0$ we will consider the position of the string to the right with projection $x + \Delta x$ and the position of the string to the left of with projection $x - \Delta x$. Certainly, the current time and even the particular we consider on this string has no effect on this force, so we expect $u_{tt}(x, t)$ to only depend on the relative position of the string itself. That is, if the displacements to the right and left of x are more positive than at x itself, then we expect a force in the positive direction. (See Figure) If the displacement to the right is positive and the displacement to the left is negative, we expect some cancelation in the forces from the right and left.

In particular, if we average the two differences (over Δx) with the string's location at x itself we get $\frac{u(x - \Delta x, t) - u(x, t)}{\Delta x}$ and $\frac{u(x + \Delta x, t) - u(x, t)}{\Delta x}$ we obtain a good

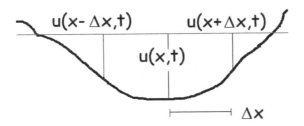

Figure 8.5: Since $u(x + \Delta x, t) > u(x, t)$ and $u(x - \Delta x, t) > u(x, t)$ we expect $F(x, t) > 0$ (an upward force).

approximation for what this force $F(x, t)$ is proportional to, namely:

$$\frac{\frac{u(x - \Delta x, t) - u(x, t)}{\Delta x} + \frac{u(x + \Delta x, t) - u(x, t)}{\Delta x}}{\Delta x},$$

or

$$F(x, t) \propto \frac{u(x - \Delta x, t) + u(x + \Delta x, t) - 2u(x, t)}{(\Delta x)^2}.$$

Taking $\Delta x \to 0$ the right hand side converges to $u_{xx}(x, t)$.

In sum, we have informally derived:

The Wave Equation

An elastic string projects perpendicularly onto the interval $[0, L]$ with perpendicular displacement $u(x, t)$. Then its motion is given by

$$u_{tt}(x, t) = \beta u_{xx}(x, t) \qquad (8.7)$$

where $\beta > 0$ is a constant called the constant of propagation.

We note some 'easy' solutions that have constant velocity which are $u(x, t) = ax + bt + c$.

If we insist on fixing the location of the ends at zero, then the boundary conditions become $u(0, t) = 0$ and $u(L, t) = 0$ (See figure).

Next we search for separable solutions to the wave equation with $u(0, t) = 0$ and $u(L, t) = 0$. Suppose $u(x, t) = X(x)T(t)$. Then $u_{xx}(x, t) = X''(x)T(t)$ and $u_{tt}(x, t) = X(x)T''(t)$ and we obtain

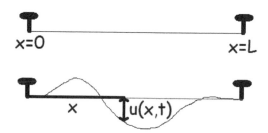

Figure 8.6: Above: An elastic string with no displacement $u(x,t) = 0$. Below: An elastic string with displacement over a superimposed number line. Both have fixed homogeneous ends.

$X(x)T''(t) = \beta X''(x)T(t)$ and separating, we get

$$\frac{X''(x)}{X(x)} = \frac{1}{\beta}\frac{T''(t)}{T(t)}.$$

The left side only depends on x and the right side only depends on t so both quotients must be constants. We set this constant equal to $-\lambda$.

$$\frac{X''(x)}{X(x)} = \frac{1}{\beta}\frac{T''(t)}{T(t)} = -\lambda.$$

At this point, we launch into the infamous three cases. We focus on the possible solutions to

$$\frac{X''(x)}{X(x)} = -\lambda.$$

Case 1: $\lambda < 0$. Note that $-\lambda > 0$.
This means

$$\frac{X''(x)}{X(x)} = -\lambda$$

implies that

$$X''(x) + \lambda X(x) = 0$$

where $-\lambda > 0$.
This has solutions

$$X(x) = c_1 e^{\sqrt{-\lambda}x} + c_2 e^{-\sqrt{-\lambda}x}$$

where $-\lambda > 0$.

Since $X(0) = 0$ we see that $c_1 + c_2 = 0$ and $X(L) = 0$ yields $c_1 e^{\sqrt{-\lambda}L} + c_2 e^{-\sqrt{-\lambda}L} = 0$.

Since $c_1 = -c_2$ we see that

$$-c_2 e^{\sqrt{-\lambda}L} + c_2 e^{-\sqrt{-\lambda}L} = 0$$

or

$$c_2(e^{-\sqrt{-\lambda}L} - e^{\sqrt{-\lambda}L}) = 0$$

The term in the parentheses is zero only when $\lambda = 0$ or $L = 0$ so we must have $c_1 = c_2 = 0$, so this means the only possible separable solution with $\lambda < 0$ is $X(x) = 0$ so $u(x, t) = 0$.

Case 2: $\lambda = 0$. In this case,

This means

$$\frac{X''(x)}{X(x)} = 0$$

implies that

$$X''(x) = 0.$$

This has solutions

$$X(x) = c_1 + c_2 x$$

Since $X(0) = 0$ we see that $c_1 = 0$ and $X(L) = 0$ yields $c_1 + c_2 L = 0$, which forces $c_2 = 0$.

This means the only possible separable solution with $\lambda = 0$ is $X(x) = 0$ so $u(x, t) = 0$.

Case 3: $\lambda > 0$. In this case,

This means

$$\frac{X''(x)}{X(x)} = -\lambda$$

implies that

$$X''(x) + \lambda X(x) = 0.$$

This has solutions

$$X(x) = c_1 \cos(\sqrt{\lambda}x) + c_2 \sin(\sqrt{\lambda}x)$$

Since $X(0) = 0$ we see that $c_1 = 0$ and $X(L) = 0$ yields $0 = X(L) = c_2 \sin(\sqrt{\lambda}L)$.

This has solution $c_2 = 0$ (which again lead to $u(x,t) = 0$) or $\sqrt{\lambda}L = k\pi$ where k is a positive integer.

Therefore when

$$\lambda = \frac{k^2\pi^2}{L^2},$$

we have a solution which satisfies $X(0) = 0$ and $X(L) = 0$. This solution is

$$X(x) = c_2 \sin(\frac{k\pi}{L}x)$$

Now we find the associated solution to

$$\frac{T''(t)}{\beta T(t)} = -\lambda = -\frac{k^2\pi^2}{L^2}$$

or

$$T''(t) + \frac{\beta k^2\pi^2}{L^2}T(t) = 0.$$

The solutions are would be $T(t) = c_1 \cos(\frac{\sqrt{\beta}k\pi}{L}t) + c_2 \sin(\sqrt{\beta}\frac{k\pi}{L}t)$.

Thus there are infinitely many separable solutions of the form

$$u_k(x,t) = \sin(\frac{k\pi}{L}x)\left[A_k \cos(\frac{\sqrt{\beta}k\pi}{L}t) + B_k \sin(\frac{\sqrt{\beta}k\pi}{L}t)\right]$$

for $k = 1, 2, 3, ...$ together with $u_0(x,t) = 0$.

Taking all possible linear combinations (since the boundary conditions are zero, linear combinations of these solutions will still satisfy the homogeneous boundary condition).

The Wave Equation

Solutions to

$$u_{tt}(x,t) = \beta u_{xx}(x,t) \tag{8.8}$$

where $\beta > 0$ is a constant called the constant of propagation with $u(0,t) = 0$ and $u(L,t) = 0$ are given by

$$u(x,t) = \sum_{k=1}^{\infty} \sin(\frac{k\pi}{L}x) \left[A_k \cos(\frac{\sqrt{\beta}k\pi}{L}t) + B_k \sin(\frac{\sqrt{\beta}k\pi}{L}t) \right]$$

where

$$u(x,0) = \sum_{k=1}^{\infty} A_k \sin(\frac{k\pi}{L}x)$$

$$u_t(x,0) = \sum_{k=1}^{\infty} B_k \frac{\sqrt{\beta}k\pi}{L} \sin(\frac{k\pi}{L}x)$$

Here A_k are the Fourier coefficients of the odd extension of $u(x,0)$ to $[-L,L]$ and $B_k = \frac{L}{\sqrt{\beta}k\pi}C_k$, where C_k are the Fourier coefficients of the odd extension of $u_t(x,0)$ to $[-L,L]$.

$\square$

As the above result indicates, the motion of the wave depends on the initial displacement given by $u(x,0)$ and the initial velocity at each location given by $u_t(x,0)$. Effectively, the A_k are the coefficients of the Fourier sine series of $u(x,0)$ on $[-L,L]$ and the B_k are obtained taking the coefficients of the Fourier sine series of $u_t(x,0)$ on $[-L,L]$ and scaling them by $\frac{L}{\sqrt{\beta}k\pi}$. Furthermore, the solutions will continue to oscillate forever. As in the undamped spring mass system, there is no energy loss in this system.

Example 8.3 *Consider the wave equation with $\beta = 1$ and $L = 3$. The ends are held at a constant with $u(0,t) = 0$ and $u(3,t) = 0$. The initial displacement is $x(x-1)(x-3)$ on $(0,3)$ and the initial velocity is initially zero at all points. Find $u(x,t)$ and plot approximations for $t = 0, 0.01, 0.1,$ and 1.*

Solution: Since the initial velocity is zero, $u_t(x,0) = 0$ and so all $B_k = 0$. To find the A_k we compute the Fourier sine series of $x(x-1)(x-3)$ on $(0,3)$ and obtain:

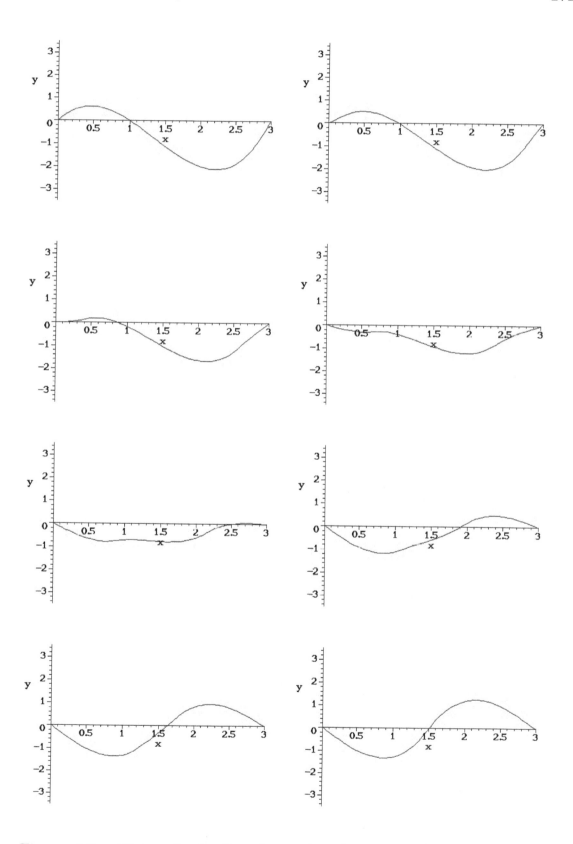

Figure 8.7: Plots of $u(x,0)$, $u(x,0.21)$, $u(x,0.42)$, $u(x,0.63)$, $u(x,0.84)$, $u(x,1.05)$, $u(x,1.26)$, $u(x,1.47)$ and $u(x,1.68)$ for Example 8.3

$A_k = \frac{144+(-1)^n 180}{k^3 \pi^3}$ and we plot $u(x,t)$ for $t = 0, 0.21, 0.42, 0.63, 0.84, 1.05, 1.26,$ 1.47, and 1.68.

Example 8.4 *Consider the wave equation with $\beta = 1$ and $L = 3$. The ends are held at a constant with $u(0,t) = 0$ and $u(3,t) = 0$. The initial displacement is $x(x-1)(x-3)$ on $(0,3)$ and the initial velocity is initially equal to x at all points. Find $u(x,t)$ and plot approximations for $t = 0, 0.01, 0.1,$ and 1.*

Solution: The A_k are as in the previous example $A_k = \frac{144+(-1)^n 180}{k^3 \pi^3}$. To obtain the B_k, we first compute the Fourier coefficients of $g(x) = x$ extended as an odd function on $(-3,3)$. We obtain $C_k = 6\frac{(-1)^{n+1}}{n\pi}$. Next we obtain the B_k by scaling these by $\frac{3}{n\pi}$ to obtain $B_k = 18\frac{(-1)^{n+1}}{n^2\pi^2}$ We plot $u(x,t)$ for $t = 0, 0.21, 0.42, 0.63, 0.84, 1.05, 1.26, 1.47,$ and 1.68.

$\square$

Nonhomogeneous ends

As before we can consider the case where the position at the ends is not zero by subtracting the steady state solution.

For instance if the end of at zero is to be held constant at A and the end at L is to be held constant at B, then once again we consider the solution

$$u_{ss}(x,t) = \left(\frac{B-A}{L}\right)x + A.$$

Then for any solution to this problem,

$u_{homog}(x,t) = u(x,t) - u_{ss}(x,t)$ will solve the wave equation with homogeneous ends.

We solve for the correct $u_{homog}(x,t)$ by realizing that

$$v(x,0) = u(x,0) - \left(\left(\frac{B-A}{L}\right)x + A\right)$$

and that $(u_{homog})_t(x,0) = u_t(x,0)$.

In other words:

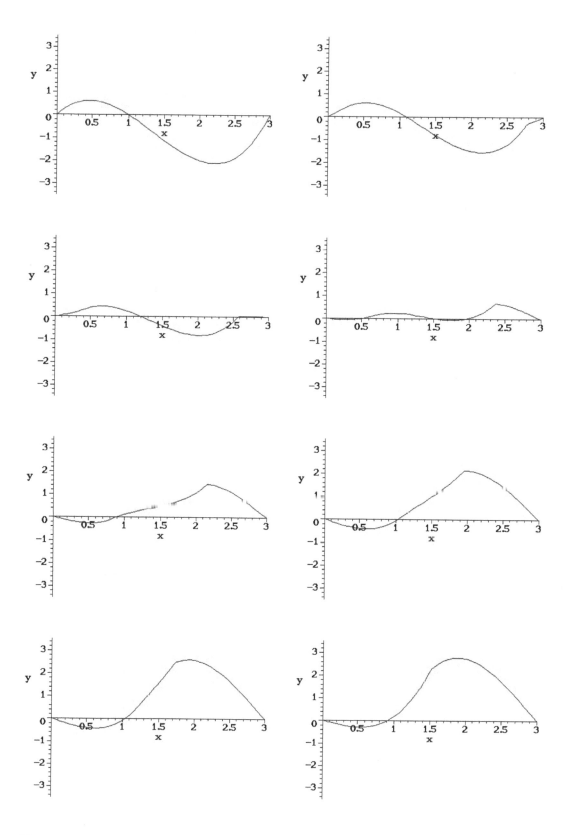

Figure 8.8: Plots of $u(x, 0)$, $u(x, 0.21)$, $u(x, 0.42)$, $u(x, 0.63)$, $u(x, 0.84)$, $u(x, 1.05)$, $u(x, 1.26)$, $u(x, 1.47)$ and $u(x, 1.68)$ for Example 8.4

Wave Equation with Constant Nonhomogeneous Ends

The heat equation

$$u_{tt} = \beta u_{xx} \tag{8.9}$$

with conditions $u(0,t) = A$ and $u(L,t) = B$ has solutions

$$u(x,t) = \sum_{k=1}^{\infty} \sin(\frac{k\pi}{L}x) \left[A_k \cos(\frac{\sqrt{\beta}k\pi}{L}t) + B_k \sin(\frac{\sqrt{\beta}k\pi}{L}t) \right]) + \left(\frac{B-A}{L} \right) x + A,$$

moreover, if the initial displacement is is $u(x,0)$, then the constants A_k are the Fourier coefficients of

$$u(x,0) - \left(\frac{B-A}{L} \right) x - A$$

expanded as an ODD function on $[-L,L]$ and the B_k are obtained taking the coefficients of the Fourier of $u_t(x,0)$ extended as an ODD function on $[-L,L]$ and scaling them by $\frac{L}{\sqrt{\beta}k\pi}$.

Example 8.5 *Consider the wave equation with $\beta = 2$ and $L = 3$. The ends are held at a constant with $u(0,t) = 2$ and $u(3,t) = 4$. The initial displacement is $u(x,0) = x + 1$ on $(0,3)$ and the initial velocity is initially equal to x at all points.*

Solution: First, $u_{SS}(x) = \frac{2}{3}x + 2$.

So we need to find the Fourier odd coefficients of
$u_{homog}(x,0) = x + 1 - (\frac{2}{3}x + 2) = \frac{1}{3}x - 1$ where $L = 3$. which is $A_j = -\frac{2}{\pi j}$
and of
$(u_{homog})_t(x,0) = u_t(x,0) = x$
which is $C_j = -\frac{6(-1)^j}{\pi j}$. To get the constants B_j we multiply by $\frac{L}{\sqrt{\beta}j\pi}$, so
$B_j = -\frac{18(-1)^j}{\sqrt{2}\pi^2 j^2}$.

Thus,

$$u(x,t) = \sum_{j=1}^{\infty} \sin(\frac{1}{3}j\pi x) \left[\left(-\frac{2}{\pi j} \right) \cos(\frac{\sqrt{2}}{3}j\pi t) + \left(-\frac{18(-1)^j}{\sqrt{2}\pi^2 j^2} \right) \sin(\frac{\sqrt{2}}{3}j\pi t) \right] + \frac{2}{3}x + 2$$

$\square$

Exercises

1. Find the solution to the wave equation with $\beta = 1$, $L = 1$ and $u(0, t) = 0$, $u(1, t) = 0$ having initial position $u(x, 0) = \sin(\pi x)$ and zero initial velocity. Plot $u(x, t)$ for several fixed values of t.

2. Find the solution to the wave equation with $\beta = 1$, $L = 1$ and $u(0, t) = 0$, $u(1, t) = 0$ having initial position $u(x, 0) = 1$ and zero initial velocity. Plot $u(x, t)$ for several fixed values of t.

3. Find the solution to the wave equation with $\beta = 1$, $L = 1$ and $u(0, t) = 0$, $u(1, t) = 0$ having initial velocity $u_t(x, 0) = 1$ and no initial displacement. Plot $u(x, t)$ for several fixed values of t.

4. Find the solution to the wave equation with $\beta = 1$, $L = 1$ and $u(0, t) = 0$, $u(1, t) = 0$ having initial position $u(x, 0) = x$ and initial velocity $u_t(x, 0) = 1$. Plot $u(x, t)$ for several fixed values of t.

5. Find the solution to the wave equation with $\beta = 1$, $L = 1$ and $u(0, t) = 2$, $u(1, t) = 5$ having initial position $u(x, 0) = 3x + 2$ and initial velocity $u_t(x, 0) = (x - 2)(x - 5)$. Plot $u(x, t)$ for several fixed values of t.

6. Find the solution to the wave equation with $\beta = 1$, $L = 2$ and $u(0, t) = 1$, $u(1, t) = 2$ having initial position $u(x, 0) = x^2 + 1$ and initial velocity $u_t(x, 0) = 1$. Plot $u(x, t)$ for several fixed values of t.

8.3   Laplace Equation

In this section, we consider the temperature $u(x, y)$ at location (x, y), where we restrict our attention to the rectangle $0 \leq x \leq L$ and $0 \leq y \leq M$. It turns out that the steady state solutions to the heat equation on this rectangle solves the partial differential equation:

Laplace Equation

The steady state solution of the heat equation on a rectangle $0 \leq x \leq L$ and $0 \leq y \leq M$, where

$u(x, y)$ is the temperature at location (x, y) is given by

$$u_{xx}(x, y) + u_{yy}(x, y) = 0 \qquad (8.10)$$

Specifically, the solutions to Laplace's equation are the functions whose gradients are divergent free. That is, these functions have the property that their gradient vector fields preserve area. These functions appear in many applications, such as incompressible fluid flow and electrostatics.

As usual, we start with homogeneous initial conditions: $u(0, y) = 0$, $u(L, y) = 0$, $u(x, M) = 0$ and allow $u(x, 0)$ to be arbitrary. (If we set $u(x, 0) = 0$, we would have the trivial solution.) We all know what comes next. We search for separable solutions to this equation with $u(0, y) = 0$, $u(L, y) = 0$, and $u(x, M) = 0$. Suppose $u(x, y) = X(x)Y(y)$. Then $u_{xx}(x, y) = X''(x)Y(y)$ and $u_{yy}(x, y) = X(x)Y''(y)$ and we obtain $X(x)Y''(y) + X''(x)Y(y) = 0$ and separating, we get

$$\frac{X''(x)}{X(x)} = -\frac{Y''(t)}{Y(y)}.$$

The left side only depends on x and the right side only depends on y so both quotients must be constants. We set this constant equal to $-\lambda$.

$$\frac{X''(x)}{X(x)} = -\frac{Y''(t)}{Y(y)} = -\lambda.$$

We launch into the infamous three cases. We focus on the possible solutions to

$$\frac{X''(x)}{X(x)} = -\lambda.$$

Case 1: $\lambda < 0$. Note that $-\lambda > 0$.

This means

$$\frac{X''(x)}{X(x)} = -\lambda$$

implies that

$$X''(x) + \lambda X(x) = 0$$

where $\lambda < 0$.

This has solutions

$$X(x) = c_1 e^{\sqrt{-\lambda}x} + c_2 e^{-\sqrt{-\lambda}x}$$

where $-\lambda > 0$.

Since $X(0) = 0$ we see that $c_1 + c_2 = 0$ and $X(L) = 0$ yields $c_1 e^{\sqrt{-\lambda}L} + c_2 e^{-\sqrt{-\lambda}L} = 0$.

Since $c_1 = -c_2$ we see that

$$-c_2 e^{\sqrt{-\lambda}L} + c_2 e^{-\sqrt{-\lambda}L} = 0$$

or

$$c_2(e^{-\sqrt{-\lambda}L} - e^{\sqrt{-\lambda}L}) = 0$$

The term in the parentheses is zero only when $-\lambda = 0$ or $L = 0$ so we must have $c_1 = c_2 = 0$, so this means the only possible separable solution with $\lambda < 0$ is $X(x) = 0$ so $u(x, y) = 0$.

Case 2: $\lambda = 0$. In this case,

This means

$$\frac{X''(x)}{X(x)} = 0$$

implies that

$$X''(x) = 0.$$

This has solutions

$$X(x) = c_1 + c_2 x$$

Since $X(0) = 0$ we see that $c_1 = 0$ and $X(L) = 0$ yields $c_1 + c_2 L = 0$, which forces $c_2 = 0$.

This means the only possible separable solution with $\lambda = 0$ is $X(x) = 0$ so $u(x, y) = 0$.

Case 3: $\lambda > 0$. In this case,

This means

$$\frac{X''(x)}{X(x)} = -\lambda$$

implies that

$$X''(x) + \lambda X(x) = 0.$$

This has solutions

$$X(x) = c_1 \cos(\sqrt{\lambda}x) + c_2 \sin(\sqrt{\lambda}x)$$

Since $X(0) = 0$ we see that $c_1 = 0$ and $X(L) = 0$ yields $0 = X(L) = c_2 \sin(\sqrt{\lambda}L)$.

This has solution $c_2 = 0$ (which again lead to $u(x,t) = 0$) or $\sqrt{\lambda}L = k\pi$ where k is a positive integer.

Therefore when

$$\lambda = \frac{k^2\pi^2}{L^2},$$

we have a solution which satisfies $X(0) = 0$ and $X(L) = 0$. This solution is

$$X(x) = c_2 \sin(\frac{k\pi}{L}x)$$

Now we find the associated solution to

$$\frac{Y''(y)}{Y(t)} = \lambda$$

where $\lambda = \frac{k^2\pi^2}{L^2}$.

We obtain

$$Y''(y) - \frac{k^2\pi^2}{L^2}Y(y) = 0.$$

The solutions are $Y(y) = c_1 e^{\frac{k\pi}{L}y} + c_2 e^{-\frac{k\pi}{L}y}$.

Note that $Y(M) = 0$ so we have

$0 = c_1 e^{\frac{k\pi}{L}M} + c_2 e^{-\frac{k\pi}{L}M}$.

which implies that: $c_2 = -c_1 e^{\frac{2k\pi}{L}M}$. So $Y(y) = c_1 e^{\frac{k\pi}{L}y} - c_1 e^{\frac{2k\pi}{L}M}e^{-\frac{k\pi}{L}y}$ or $Y(y) = c_1\left(e^{\frac{k\pi}{L}y} - e^{\frac{k\pi}{L}(2M-y)}\right)$ After relabelling the constants, we see that there are infinitely many separable solutions of the form

$$u_k(x,y) = c_k \sin(\frac{k\pi}{L}x)\left[e^{\frac{k\pi}{L}y} - e^{\frac{k\pi}{L}(2M-y)}\right]$$

for $k = 1, 2, 3, \ldots$ together with $u_0(x,y) = 0$. Noting that linear combinations of solutions are also solutions, we obtain the following:

Laplace Equation on a Rectangle with Three Constant Homogeneous Ends
The equation

$$u_{xx} + u_{yy} = 0 \tag{8.11}$$

with boundary conditions $u(0,y) = 0$, $u(L,y) = 0$, and $u(x,M) = 0$. has solutions

$$u(x,y) = \sum_{k=1}^{\infty} c_k \sin(\frac{k\pi}{L}x)\left[e^{\frac{k\pi}{L}y} - e^{\frac{k\pi}{L}(2M-y)} \right],$$

moreover, if the initial condition is specified as $u(x,0) = f(x)$, then the constants $c_k\left(1 - e^{\frac{2k\pi M}{L}}\right)$ are the odd Fourier coefficients of $f(x)$ on $[0,L]$.

Example 8.6 *Solve the Laplace equation with $L = 1$ and $M = 2$ with $u(x,0) = x(1-x)$. Plot the surface $z = u(x,y)$ in three dimensions for $0 \le x \le 1$ and $0 \le y \le 2$.*

Solution: We first find the Fourier odd coefficients of $f(x) = x(1-x)$. These are (after 2 iterations of integration by parts)

$$d_k = \frac{4 - \cos(k\pi)}{k^3\pi^3}$$

Thus

$$c_k = d_k \frac{1}{(1 - e^{4\pi k})} = \frac{4 - \cos(k\pi)}{k^3\pi^3 (1 - c^{4\pi k})}$$

Therefore,

$$u(x,y) = \sum_{k=1}^{\infty} \frac{4 - \cos(k\pi)}{k^3\pi^3 (1 - e^{4\pi k})} \sin(\frac{k\pi}{x})\left[e^{k\pi y} - e^{k\pi(4-y)} \right],$$

A plot of the surface $z = u(x,y)$ is shown below. Note that $u(x,0) = x(1-x)$ and the function is 0 on the other 3 edges of the rectangular domain.

$\square$

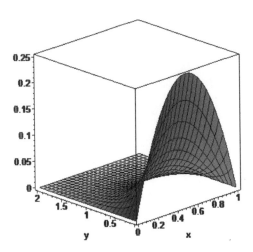

Figure 8.9: A solution to the Laplace equation.

Exercises

1. Find the solution to Laplace's equation with $L = 1$, $M = 2$, with boundary conditions $u(0, y) = 0$, $u(L, y) = 0$, $u(x, M) = 0$, and $u(x, 0) = \sin(2\pi x)$,

   Plot $u(x, y)$ as a surface.

2. Find the solution to Laplace's equation with $L = 1$, $M = 2$ with boundary conditions $u(0, y) = 0$, $u(L, y) = 0$, $u(x, M) = 0$, and $u(x, 0) = x^2(1 - x)$, Plot $u(x, y)$ as a surface.

3. Find the solution to Laplace's equation with $L = 1$, $M = 3$ with boundary conditions $u(0, y) = 0$, $u(L, y) = 0$, $u(x, M) = 0$, and $u(x, 0) = 1 - |x|$, Plot $u(x, y)$ as a surface.

4. Find the solution to Laplace's equation with $L = 3$, $M = 3$ with boundary conditions $u(0, y) = 0$, $u(L, y) = 0$, $u(x, M) = 0$, and $u(x, 0) = x$, Plot $u(x, y)$ as a surface.

Solutions to Selected Problems
Section 1.1

1. This is a first order ODE, function z, independent variable t.

2. This is a third order PDE, function u, independent variables x, y, z.

3. This is a third order ODE, function y, independent variable x.

4. This is a first order ODE, function y, independent variable x (or the reverse, since it is written as a differential).

5. This is a first order PDE, function z, independent variables x, y.

6. This is a second order ODE, function y, independent variables x.

9. For $y = \sqrt{x}$ we have $y'' = -\frac{1}{4}y^{-\frac{3}{2}}$. We plug this into the DE and see that $x^2\left(-\frac{1}{4}y^{-\frac{3}{2}}\right) + \frac{1}{4}(\sqrt{x})$ and that this is indeed equal to zero. So $y = \sqrt{x}$ is a solution to the ODE.

11. For $x^3 + y^4 = xy$ we implicitly differentiate to get $3x^2 + 4y^3\frac{dy}{dx} = x\frac{dy}{dx} + y$ which simplifies to $\frac{dy}{dx} = \frac{y-3x^2}{4y^3-x}$. This does NOT agree with the DEs in (a) and (c) but DOES agree with the DE in (b). So the equation solves (b).

12. $\frac{dP}{dt} = kP(t)$

13. $\frac{dM}{dt} = -2\frac{M(t)}{400-2t}$

16. (a) $y = \frac{1}{2}\arctan(\frac{x}{2}) + \pi - \frac{1}{2}\arctan(\frac{1}{2})$

Section 2.1

1. $z = Ce^{\frac{t^2}{2}}$

3. $y = \tan(\frac{x^2}{2} + C)$

5. $z = -\frac{1}{t+c}$ and $z = 0$

7. $y = \sqrt[3]{-\frac{3x^2}{2} + C}$ and $y = 0$

10. $y = \left(\frac{x+3}{2}\right)^2$

13. $r = \ln|\theta| + 1 - \ln \pi$

Section 2.2

1. $y = Cx + \frac{1}{3}x^4$

3. $z = Ce^t - t - 1$

5. $z = \frac{1}{2}\sin t + \frac{C}{\sin t}$

8. $y = 2 - 3e^{-x}$

9. $z = \frac{1}{2} + \frac{3e}{2e^{x^2}} = \frac{1}{2} + \frac{3}{2}e^{1-x^2}$

Section 2.3

1. $M_y = 2y$, $\ N_x = 1$, so the DE is not exact.

3. $M_s = 1$, $\ N_r = 1$, so the DE is exact.

5. $x\sqrt{y} + x^2 \tan y = C$

8. $x\sqrt{x^2 + y^2} = C$

10. $x^2 y^3 = 8$

12. (a) $M_y = 4xy^3$, $\ N_x = 8xy^3$, so the DE is not exact.

    (b) $M_y = 4y^3$, $\ N_x = 4y^3$, so the DE is exact.

    (c) $xy^4 = C$ is the solution. It solves the original DE (check by implicit differentiation, then multiply the resulting equation by x).

Section 2.4

2. Between $t_1 = -\frac{5730\ln(0.15)}{\ln 2} \approx 15682.8$ and $t_2 = -\frac{5730\ln(0.35)}{\ln 2} \approx 8678.5$ years old.

3. $t = -\frac{1600\ln(\frac{1}{1881})}{\ln 2} \approx 17403.65$ years.

6. Using $r = .05$, and initial conditions, we get $Q = \frac{K}{.05} + (100000 - \frac{K}{.05})e^{.05t}$

   So solving $Q(30) = 0$ for K we get $K = \frac{100000e^{1.5}}{1-e^{1.5}}.05 \approx \$6436.08$. This is an annual payment rate since t is in years.

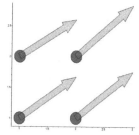

Slope field for 2.7 # 1

7. $t = \frac{\ln(\frac{64.6}{8})}{\ln 0.5} \approx -3.013$ so about 3 hours ago.

9. $t = \frac{\ln(\frac{12}{32})}{\ln \frac{29}{32}} \approx 9.96$ minutes.

Section 2.5

1. $\frac{ds}{dt} = 12 - \frac{s}{50+t}$. So $s(t) = \frac{6t^2 + 600t + 3750}{50+t}$ and $s(50) = 487.5$ grams.

5. $k = -\frac{1}{500} \ln(\frac{7}{12})$ and $t \approx 200.87$ years so about 2175.

6. $k = -\frac{1}{60000} \ln(2/3)$ so $t = -6\frac{\ln(9)}{\ln(\frac{2}{3})} \approx 32.51$ months.

7. For $\frac{dP}{dt} = kP(P_\infty - P)$ we implicitly differentiate with respect to t to obtain:
$\frac{d^2P}{dt^2} = k(P_\infty - P)\frac{dP}{dt} - kP\frac{dP}{dt} = k(P_\infty - 2P)\frac{dP}{dt}$
This is zero when $P = 0, P = P_\infty$, or $P = \frac{P_\infty}{2}$.

Section 2.6

1. See table.

(x, y)	$\frac{dy}{dx}$
$(1, 1)$	$\frac{2}{3}$
$(1, 2)$	$\frac{3}{5}$
$(2, 2)$	$\frac{5}{6}$
$(2, 1)$	$\frac{3}{4}$

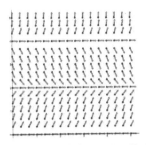

Slope field for 2.7 # 2

2. Solutions with initial y value with $0 < y < 4$ limit to $y = 2$.

3.

x_i	y_i	$f(x_i, y_i)$
0	1	1.414213562
0.2	1.282842712	1.663035004
0.4	1.615449713	1.905491912
0.6	1.996548096	2.143150996
0.8	2.425178295	2.377047873
1	2.900587869	2.60790639
1.2	3.422169147	2.836254272
1.4	3.989420002	3.062489184
1.6	4.601917839	3.286918873
1.8	5.259301613	3.509786778
2	5.961258969	3.731289045

So $y(2) \approx 5.961258969$

4.

x_i	y_i	$f(x_i, y_i)$
0	2	−6
1	−4	−60
2	−64	−12480
3	−12544	−472093440
4	−472105984	$-6.68652E + 17$

Basically Δx is too big. We know (from the slope field and since this is a logistic DE, for which we know the solutions) that $y(2)$ should be positive when $y(0) = 2$ and that this solution should limit to $y = 1$, but from our table we have $y(2) < 0$. Basically our step-size is too big.

Section 3.2

1. $y(t) = c_1 e^{-2t} + c_2 t e^{-2t}$

3. $y(t) = c_1 e^{-7t} + c_2 e^t$

5. $z(t) = c_1 e^{-t} + c_2 e^t$

7. $y(t) = c_1 e^{-2t} \cos(\sqrt{2}t) + c_2 e^{-2t} \sin(\sqrt{2}t)$

9. $y(t) = c_1 e^{\sqrt{7}t} + c_2 e^{-\sqrt{7}t}$

11. $z(t) = c_1 e^{-\frac{1}{2}t} \cos(\frac{\sqrt{3}}{2}t) + c_2 e^{-\frac{1}{2}t} \sin(\frac{\sqrt{3}}{2}t)$

13. $y(t) = e^{-t}$

15. $y(t) = e^{-4t} \sin t$

17. $y(t) = 4\cos(4t) - \frac{1}{2}\sin(4t)$

Section 3.3

1. $y(t) = \sqrt{8}\cos(t + \frac{\pi}{4})$

3. $y(t) = \cos(4t - 0)$

5. $\frac{2\pi}{\beta} = \frac{2\pi}{2}$

7. Period $= \frac{\pi}{2}$ and Amplitude $= \frac{1}{2}$

Section 3.4

1. $y(t) = \frac{3}{4}e^{-t} - \frac{3}{4}e^{-5t}$ Has a max when $y'(t) = 0$ that is when $y'(t) = -\frac{3}{4}e^{-t} + \frac{15}{4}e^{-5t} = 0$ which is when $t = \frac{1}{4}\ln 5$.

    Plugging this t-value into $y(t)$ we get maximum displacement of $\frac{3}{4}5^{-\frac{1}{4}} - \frac{3}{4}5^{-\frac{5}{4}} \approx 0.40124418298$

3. If the spring is already moving to the right and is positioned to the right, then it must eventually have a maximum value to the right (since it must limit to zero). However, it then cannot go to the left of equilibrium, since then (by similar reasons) it must again have a maxima. Overdamped, spring mass systems have at most one such maxima.

285

5. Since $y(t) = C_1 e^{(-2+\sqrt{3})t} + C_2 e^{(-2-\sqrt{3})t}$ we must have $C_1 + C_2 = -1$. By the formula (3.7) given for an equilibria $\frac{C_2}{C_1} > 1$, so since $-C_2 = 1 + C_1$ we have $\frac{1}{C_1} + 1 > 0$ or $C_1 > 0$ and so $C_2 < -1$. Thus $y'(0) = C_1(-2 + \sqrt{3}) + C_2(-2 - \sqrt{3}) = C_1(-2 + \sqrt{3}) + (-1 - C_1)(-2 - \sqrt{3}) = 2 + \sqrt{3} + 2C_1\sqrt{3}$ so if $y'(0) > 2 + \sqrt{3}$ then $C_1 > 0$.

7. $y(t) = \cos(3t)$

Section 3.5

2. $y(t) = \frac{1}{2}t^2 - \frac{7}{2}t + \frac{19}{4}$

3. $y(t) = -\frac{23}{1105}\cos(4t) + \frac{24}{1105}\sin(4t)$

5. $y(x) = 3xe^x$

7. $z(t) = \frac{7}{4}te^{-t}$

16. $z(t) = c_1 + c_2 e^{-t} + 6t$

17. $z(t) = c_1 e^{-6t} + c_2 e^{-t} + \frac{1}{5}te^{-t}$

20. $z(t) = \frac{7}{9}e^{-2t} + \frac{4}{3}te^{-2t} + \frac{2}{9}e^t$

Section 3.6

1. For constant forcing $y_{SS}(t) = \frac{4}{10}$, for the other forcing, $y_{SS}(t) = \frac{6}{13}\cos(2t) + \frac{4}{13}\sin(2t)$ which has amplitude $\sqrt{\left(\frac{6}{13}\right)^2 + \left(\frac{4}{13}\right)^2} \approx 0.5547$. Also, since $2mk - b^2 < 0$ this system exhibits resonance.

4. (a) $y_{SS}(t) = 1$

   (c) For $\omega = \frac{1}{\sqrt{2}}$ we have $y_{SS}(t) = \frac{4}{3}\cos(\frac{1}{\sqrt{2}}t) + \frac{4\sqrt{2}}{3}\sin(\frac{1}{\sqrt{2}}t)$ which has amplitude $\sqrt{\left(\frac{4}{3}\right)^2 + \left(\frac{4\sqrt{2}}{3}\right)^2} = \frac{\sqrt{32}}{3}$.

6. $\omega_{res} = \frac{1}{\sqrt{2}} = \frac{\sqrt{2}}{2}$. So after undetermined coefficients, $y_{SS}(t) = \frac{2}{3}\cos(\frac{1}{\sqrt{2}}t) + \frac{2\sqrt{2}}{3}\sin(\frac{1}{\sqrt{2}}t)$, which has amplitude $\frac{2\sqrt{3}}{3} \approx 1.15$

Section 3.7

3. $q(t) = 5e^{-\frac{1}{6}t}\cos(\frac{\sqrt{2}}{12}t) + 17\sqrt{2}e^{-\frac{1}{6}t}\sin(\frac{\sqrt{2}}{12}t)$

5. $q_{ss}(t) = 1$

6. Overdamped: $R^2 - 4\frac{L}{C} > 0$,

   Critically damped: $R^2 - 4\frac{L}{C} = 0$,

   Underdamped: $R^2 - 4\frac{L}{C} < 0$,

   Lightly underdamped: $R^2 - 2\frac{L}{C} < 0$,

   Resonance (for maximizing charge): $\omega_{res} = \frac{\sqrt{2\frac{L}{C} - R^2}}{\sqrt{2}L}$.

Section 3.8

2. $\frac{1}{2}t - \frac{3}{4}$

3. $-\ln|\sec t + \tan t|\cos t$

4. $y(t) = -t\cos t + \ln|\sin t|\sin t$

7. $y(t) = c_1\cos t + c_2\sin t - t\cos t + \ln|\sin t|\sin t$

11. $y(t) = c_1\cos 3t + c_2\sin 3t - (-\frac{1}{10}\cos(5t) - \frac{1}{2}\cos(t))cos(3t) + (\frac{1}{2}\sin(t) + \frac{1}{10}\sin(5t))\sin(3t)$

    Where (after technology) $c_1 = \frac{6}{5}\cos(1)\sin^2(1) - \frac{8}{5}\cos(1)\sin^4(1) - \frac{3}{5}\cos(1) - 2\sin(1) + \frac{8}{3}\sin^3(1)$,

    and $c_2 = -\frac{8}{5}\sin^5(1) + 2\sin^3(1) - \frac{8}{3}\cos(1)\sin^2(1) - sin(1) + \frac{2}{3}\cos(1)$.

13. $y(t) = c_1\cos t + c_2\sin t - (\int_0^t \frac{\sin^2 w}{w}\,dw)\cos t + (\int_0^t \frac{\cos w\sin w}{w}\,dw)\sin t$

    Here $y(0) = 0$ implies $c_1 = 0$ since the integrals are zero and $y'(0) = 1$ implies (after differentiating the messy expression above and evaluating at $t = 0$) that $c_2 = 1$

Section 4.4

1. $y(t) = c_1 + c_2e^{-t} + c_3te^{-t}$

3. $y(t) = c_1e^{-t} + c_2e^{-2t} + c_3e^{-3t}$

5. $y(t) = c_1\cos t + c_2\sin t + c_3\cos(2t) + c_4\sin(2t)$

7. $y(t) = \frac{1}{2}e^{2t} + \frac{1}{2}\cos(2t) + \frac{1}{2}\sin(2t)$

Section 5.1

1. $\mathcal{L}[0] = 0$

3. $\dfrac{1}{s^2}\left(1 - 2e^{-s} + e^{-2s}\right),\ s > 0$

5. $\mathcal{L}[\cos(\beta t)] = \frac{s}{s^2 + \beta^2},\ s > 0$

7. $-\dfrac{8}{s} + \dfrac{1}{s^2} - \dfrac{e^{-8s}}{s^2} + \dfrac{e^{-8s+14} - e^{-10s+16}}{s-1} + \dfrac{(-121s^2 - 22s - 2)e^{-11s}}{s^3} + \dfrac{(100s^2 + 20s + 2)e^{-10s}}{s^3},$
$s > 0$

11. $\dfrac{1}{s} + \dfrac{35}{s^2 + 25}$

Section 5.2

2. $\frac{1}{6}t^3$

3. $\cos(4t) + \frac{5}{4}\sin(4t)$

6. $\frac{1}{2}\cos t + \frac{1}{2}\sin t - \frac{1}{2}e^{-t}$

8. $y = c_1 e^{-t} + c_2 e^{-5t} + \frac{1}{5}t - \frac{6}{25}$

9. $y = c_1 e^{-t} + c_2 e^{t} + \frac{1}{2}te^{t}$

11. $y = -\cos t - \sin t + 2$

Section 5.3

2. $y = \frac{25}{27}e^{-t}\cos(\sqrt{2}t) + \frac{37}{54}\sqrt{2}e^{-t}\sin(\sqrt{2}t) + \frac{1}{3}t^3 - \frac{4}{9}t + \frac{2}{27}$

3. $y = \frac{2}{39}e^{-3t}\cos t + \frac{3}{39}e^{-3t}\sin t + \frac{1}{13}\sin t - \frac{2}{39}\cos t$

4. $y = \frac{12}{13}e^{-t}\cos(2t) + \frac{23}{26}e^{-t}\sin(2t) + \frac{1}{13}e^{t}$

6. $y = -\frac{1}{3}t^3$

1. $f(t) = 1 + (e^t - 1)u(t - 1)$

   $\mathcal{L}[f(t)] = \frac{1}{s} + e^{-s}\left[\frac{e^1}{s-1} - \frac{1}{s}\right]$

2. $f(t) = [1 - u(t - \pi)]\sin t + u(t - \pi)\cos t$

   To do $\mathcal{L}$ we use $\sin(a + b) = \sin(a)\cos(b) + \cos(a)\sin(b)$ and $\cos(a + b) = \cos(a)\cos(b) - \sin(a)\sin(b)$

   $\mathcal{L}[f(t)] = \frac{1}{s^2+1} + e^{-\pi s}\left[\frac{1}{s^1+1} - \frac{s}{s^2+1}\right]$

7. $y = \frac{1}{4}(t-2)^2 u(t-2) - \frac{3}{4}u(t-2)(t-2) - \frac{1}{2}u(t-2) - \frac{5}{4\sqrt{3}}u(t-2)e^{-t+2}\sin(\sqrt{3}(t-2)) - \frac{1}{2}u(t-2)e^{-t+2}\cos(\sqrt{3}(t-2)) - \frac{1}{4}(t-3)^2 u(t-3) - 2u(t-3)(t-3) - \frac{9}{8}u(t-3) + \frac{9}{8}u(t-3)e^{-t+3}\cos(\sqrt{3}(t-3)) + \frac{25}{8\sqrt{3}}u(t-3)e^{-t+3}\sin(\sqrt{3}(t-3))$

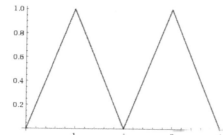

11.

12. Note: $f(t) = t - 2(t - 1)u(t - 1) + 2(t - 2)u(t - 2) - 2(t - 3)u(t - 3) + \ldots$

   So $\mathcal{L}[f(t)] = \frac{1}{s^2} - 2\frac{e^{-s}}{s^2} + 2\frac{e^{-2s}}{s^2} - 2\frac{e^{-3s}}{s^2} + \ldots$

   So $y(t) = \left[\frac{1}{4}t - \frac{1}{8} + \frac{1}{8}e^{-t}\cos(\sqrt{3}t) - \frac{1}{8\sqrt{3}}e^{-t}\sin(\sqrt{3}t)\right]$

   $-u(t-1)\left[\frac{1}{4}(t - 1) - \frac{1}{8} + \frac{1}{8}e^{-(t-1)}\cos(\sqrt{3}(t - 1)) - \frac{1}{8\sqrt{3}}e^{-(t-1)}\sin(\sqrt{3}(t - 1))\right]$

   $+u(t-2)\left[\frac{1}{4}(t - 2) - \frac{1}{8} + \frac{1}{8}e^{-(t-2)}\cos(\sqrt{3}(t - 2)) - \frac{1}{8\sqrt{3}}e^{-(t-2)}\sin(\sqrt{3}(t - 2))\right] -$

   $\ldots$ $\square$

Section 5.5

1. $\frac{1}{60}t^6$

3. $\sin t$

5. $(t-1)u(t-1)$

7. $-\frac{1}{4}t + \frac{1}{16}e^{2t} - \frac{1}{16}e^{-2t}$

9. $y(1) = \int_0^1 \frac{1}{2}\sin(2w)e^{(1-w)^2}\, dw \approx 0.44591314$

Section 5.6

1. $y(t) = u(t-1)e^{-(t-1)}(t-1) + u(t-2)e^{-(t-2)}(t-2) + u(t-3)e^{-(t-3)}(t-3)$

Section 6.2

1. $y(t) = 1 + 2t - \frac{1}{2}t^3 - \frac{1}{12}t^4 + \frac{3}{40}t^5 + \ldots$

Section 6.3

1. $y(t) = t - t^2/2 - t^3/6 + 3t^5/5! + 6t^6/6! + \ldots$

5. $y(t) = \frac{\pi}{2} - t + t^4/4! - t^5/5! - 5t^7/7! + \ldots$

Section 7.1

1. $b_5 = 4$ all other constants are zero.

3. $b_n = \frac{8(-1)^{n+1}n}{\pi(-1+2n)(1+2n)}$ (all $a_n = 0$).

   Plots shown for $4, 6, 8,$ and 100 terms.

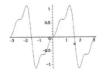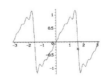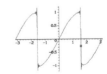

5. $a_0 = \pi$ and $a_n = 2\frac{(-1)^n - 1}{\pi n^2}$ (all $b_n = 0$).

Plots shown for $4, 6, 8,$ and 100 terms.

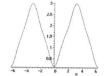 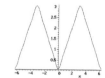 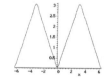 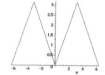

Section 7.2

1. All $b_n = 0$, $a_0 = \frac{4}{\pi}$ and $a_n = \frac{4}{\pi(4n^2-1)}$

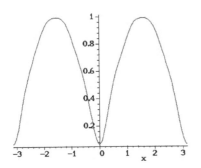
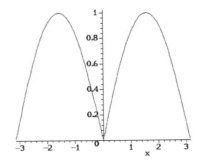

3. All $b_n = 0$, $a_0 = 5$ and $a_n = 10\left(\frac{-1+(-1)^n}{n^2\pi^2}\right)$

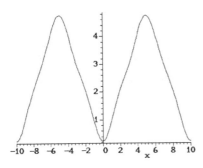
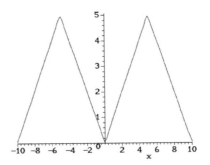

Section 8.1

1. $u(x,t) = \sin(\frac{\pi}{10}x)e^{-\frac{4}{400}\pi^2 t}$ (only $b_1 = 1$, all others are zero)

3. $u(x,t) = 5\sin(2\pi x)e^{-2\pi^2 t} - 7\sin(3\pi x)e^{-\frac{9}{2}\pi^2 t}$ (only $b_2 = 5$ and $b_3 = -7$, all others are zero)

5. $u(x,t) = \sum_{j=1}^{\infty} \frac{4}{j\pi}(-1)^{j+1}\sin(\frac{1}{2}j\pi x)e^{-\frac{1}{4}j^2\pi^2 t}$ (here $b_j = \frac{4}{j\pi}(-1)^{j+1}$, plots below are at time $t = 0, 1, 2, 3$ and we used 1000 terms in each Fourier Series)

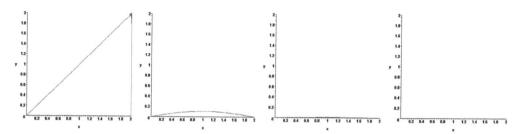

7. $u(x,t) = \sum_{j=1}^{\infty} \frac{2j\pi + j\pi \cos(2)(-1)^{j+1}}{j^2\pi^2 - 4} \sin(\frac{1}{2}j\pi x)e^{-\frac{1}{16}j^2\pi^2 t}$

9. $u_{SS}(x) = -\frac{100}{5}x + 100 = -20x + 100$

So $u_{homog}(x,0) = u(x,0) - u_{SS}(x,0) = 30 - 8\sin(6\pi x) - (-20x + 100) = 20x - 70 - 8\sin(6\pi x)$

Note that $-8\sin(6\pi x)$ is easy to solve (it is already a Fourier Series), so we just need the Fourier odd series of $20x - 70$, which is $\sum_{j=1}^{\infty} \frac{-20(7+3(-1)^j)}{j\pi} \sin(\frac{j\pi}{5}x)$
So taking both together, the solution is:

$u(x,t) = (-8)\sin(\frac{30\pi}{5}x)e^{-\frac{1}{50}(30)^2\pi^2 t} + \sum_{j=0}^{\infty}\left(\frac{-20(7+3(-1)^j)}{j\pi}\sin(\frac{j\pi}{5}x)e^{-\frac{1}{50}j^2\pi^2 t}\right) - 20x + 100$

Section 8.2

1. $u(x,t) = \sin(\pi x)\cos(\pi t)$ (only $A_1 = 1$, all others are zero.)

3. $u(x,t) = \sum_{j=1}^{\infty} \frac{1-(-1)^j}{j^2\pi^2} \sin(j\pi x)\cos(j\pi t)$ (all $A_j = 0$.)

Section 8.3

1. $u(x,y) = \frac{1}{1-e^{4\pi}} \sin(\pi x)\left[e^{\pi y} - e^{\pi(4-y)}\right]$ (only $b_1 = 1$, all others are zero.)

3. $u(x,y) = \sum_{j=1}^{\infty} \frac{1}{1-e^{6j\pi}} \frac{2}{j\pi} \sin(\pi j x)\left[e^{j\pi y} - e^{j\pi(6-y)}\right]$

293

Index

FORMULA SHEET
Laplace Transform Formulas

$\mathcal{L}[1](s) = \frac{1}{s}$

$\mathcal{L}[t^n](s) = \frac{n!}{s^{n+1}}$

$\mathcal{L}[e^{at}](s) = \frac{1}{(s-a)}$

$\mathcal{L}[\cos bt](s) = \frac{s}{s^2+b^2}$

$\mathcal{L}[\sin bt](s) = \frac{b}{s^2+b^2}$

$\mathcal{L}[\cosh bt](s) = \frac{s}{s^2-b^2}$

$\mathcal{L}[\sinh bt](s) = \frac{b}{s^2-b^2}$

$\mathcal{L}[e^{at}f(t)](s) = F(s-a)$, where $F(s) = \mathcal{L}[f(t)](s)$

$\mathcal{L}[f'(t)](s) = sF(s) - f(0)$

$\mathcal{L}[f''(t)](s) = s^2F(s) - sf(0) - f'(0)$

$\mathcal{L}[f^{(n)}(t)](s) = s^nF(s) - s^{n-1}f(0) - s^{n-2}f'(0) - \cdots - f^{(n-1)}(0)$

$\mathcal{L}[t^nf(t)](s) = (-1)^n\frac{d^n}{ds^n}(F(s))$

$\lim_{s\to\infty} F(s) = 0$

$\mathcal{L}[u(t-a)](s) = \frac{e^{-as}}{s}$ when $a > 0$

$\mathcal{L}^{-1}[e^{-as}F(s)] = f(t-a)u(t-a)$ when $a > 0$

$\mathcal{L}[f(t)u(t-a)](s) = e^{-as}\mathcal{L}[f(t+a)](s)$ when $a > 0$

$\mathcal{L}[f(t) * g(t)](s) = F(s) \cdot G(s)$

$\mathcal{L}[\delta(t-a)](s) = e^{-as}$ when $a > 0$